CHEMISTRY 105

COURSE MATERIAL

**UNIVERSITY
OF
SOUTH CAROLINA**

Department of Chemistry and Biochemistry

D. L. FREEMAN

**QDE PRESS
2016**

Copyright © 2016 by QDE Press Inc.

All rights reserved.

Permission in writing must be obtained from the publisher before any part of this work may be reproduced or transmitted in any form or by any means, electronic or mechanical, including photocopying and recording, or by any information storage or retrieval system.

Printed in the United States of America.

ISBN: 978-1-938535-01-7

QDE Press Inc.
8828 Autumnbrooke Way
Montgomery, Al 36117
www.qdepress.com

Table of Contents

Part 1 Textbook

	Page Number
Chapter 1: Chemistry and the World around You	1
Chapter 2: The Chemistry of Matter	7
Chapter 3: A Closer Look at Atoms and the Periodic Table	21
Chapter 4: Nuclear Chemistry the Chemistry of Neutrons and Protons	45
Chapter 5: Bonding of the Atoms	63
Chapter 6: Chemical Reactions	87
Chapter 7: Reactions of Acids and Bases	123
Chapter 8: The Chemistry of Carbon "The Organics"	141

Laboratory Experiments

	Page Number
Experiment 1 Safety and Laboratory Techniques	167
Experiment 2 Physical Properties of Substances	181
Experiment 3 Weight of Copper in Copper Sulfate Pentahydrate	189
Experiment 4 Energy and Chemical Reactions	199
Experiment 5 Neutralization of an Acid with a Base	209

Laboratory Experiments	**Page Number**
Experiment 6 Oxidation Reduction	221
Experiment 7 Shapes of Molecules	231
Experiment 8 Chemical Properties of Alkanes and Alkenes	241

Part 2 Lecture Slides	253
Chapter 1	Slides 1-8
Chapter 2	Slides 9-59
Chapter 3	Slides 60-155
Chapter 4	Slides 156-194
Chapter 5	Slides 195-248
Chapter 6	Slides 249-340
Chapter 7	Slides 341-394
Chapter 8	Slides 395-464

Chapter 1

Chemistry and the World Around You

Although you may not have studied chemistry in a formal setting its importance in everyday life is very obvious. The chemical reactions occurring after you eat, for example, produces energy and materials needed for normal human growth. Your clothes are made of fibers and dyes that are prepared by chemical processes. Prolonging and enhancing the quality of life through advancements in medical technology is, in a large part, associated with the chemistry of organisms and their interactions with natural and synthetic drugs. The advancements in medicinal chemistry have steadily increased since the day in 1929 when Alexander Fleming became curios as to why a fungal colony had grown on a plate of contaminated bacteria, which ultimately led to his discovery of penicillin. Mans' curiosity of how and why things interact is essential for the continued growth of humanity. Mans interest in this science extends as far back as 790,000 years; research from Hebrew and Bar-Ilan Universities in Israel where they have found evidence of controlled burning from the discovery of well-preserved hearths in cave sites. Although we have come along way since the time of early mans' knowledge of fire, there are many areas of chemistry we are just beginning to understand.

In this basic course, we will endeavor to introduce the underlying principles of chemistry and present applications of these principles to develop a clearer picture of this science and its reverence around us.

Figure 1.1
Early mans curiosity with fire probably lead to early forms of chemistry

By definition, **chemistry** is the study of matter and the changes it undergoes. But what is matter? **Matter** is anything (and I mean ANYTHING!) that has mass and occupies space. Therefore anything you can feel, smell or see[1] is matter. The air you breathe, the water you drink and this book are all examples of matter. As we will see in Chapter 2, matter can also exists as things so small that we can not see, smell, or touch them.

The second part of the definition of chemistry deals with studying the changes that matter undergoes. This process which sounds simple enough has potential to lead to many interpretations and conclusions if not carefully and scientifically performed. For instance you have heard the saying "the sun rises in the East and sets in the West". By observing the suns apparent movement you

---

[1] although you can see a shadow it is not considered matter because it does not have mass

might conclude that the sun (which, by the way, is matter) actually does revolve around the earth. This was the belief until 1543 when Copernicus published his work on the movement of heavenly bodies which scientifically proved that the earth actually revolves around the sun.

Figure 1-2.
Nicholas Copernicus'

When observing any change one must be careful not to jump to an erroneous conclusion without a careful and scientific method of analyzing the observation.

The Scientific Method

Over the centuries scientists have discovered a fail-safe method for analyzing observations; this method called the **Scientific method,** is composed of four steps:

1. Observation and description of a phenomenon or group of phenomena.

2. Formulation of a **hypothesis** (educated guess) to explain the phenomena.

3. Use of the hypothesis to predict the existence of other phenomena, or to predict the results of new observations.

4. Experimental tests of the predictions performed by several independent experimenters.

If the experimental results support the hypothesis it may become a **theory** or **law of nature**. If the experiments do not support the hypothesis, it must be rejected or modified. See Figure 1-3.

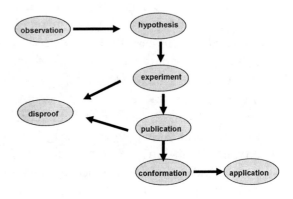

Figure 1-3.
Flow diagram of the scientific method

At some point in your life you have probably used this method to solve specific problem. For instance let's say you wake up and get ready to leave for school, however, your car won't start (Observation, step 1). You guess the battery is dead (hypothesis, step 2). You probably turn the headlights on to see how bright they are, because you know if the battery is dead the lights should be very dim (step 3), Finally, you charge the battery or get a "jump" to start your car (experiment, step 4). If your car still refuses to start, you would have to conclude that the battery is not the reason for the malfunction, requiring you to start over with a new hypothesis.

Although chemistry is considered the central science, the scientific method works for all other science disciplines. Each field retains a unique central focus but all incorporate the basic principles of matter. Science can be separated into two main categories of physical and biological. The **physical sciences** include chemistry, physics, geology, and astronomy while the **biological** sciences include botany and zoology.

Physics is the science of matter and energy and of interactions between them.

Geology is the scientific study of the origin, history, and structure of the earth.

Figure 1-4
Earth as seen from space

Figure 1-5
Jupiter and its moons

Astronomy is the scientific study of matter in outer-space, specifically the positions, dimensions, distribution, motion, composition, energy, and evolution of celestial bodies.
Botany is the science or study of plants, and **zoology** deals with animals and animal life, including the study of the structure, physiology, development, and classification of animals.

As was stated earlier, chemistry is sometimes called the "central science" because of its connection to all the other fields of study, for example when chemistry is applied to physics it is called "**physical chemistry**", when the chemistry of the earth is studied it is called "**geochemistry**", and when the chemistry of biological entities are studied it is called "**biochemistry**". Each of these disciplines have a very active role in chemical research, however, the type of research may be very different between and within each.

Types of Scientific Research

Scientific research can be separated based on the goals of the individuals engaged in the study. However, most scientific research is either considered basic science, applied science or technological science.

Basic science is research without the goal of a practical application. Benefits include contributions to culture, the possibility of discoveries of huge economic and practical importance, and broadening education. For example, research on something as common as your hand (that's right your hand, look at it). There are several different ways it can move, and it's made up of many different muscles and nerve types.

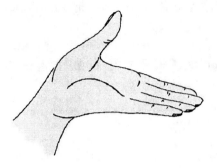

Figure 1-6
Human hand

So, basic research in how the hand works could lead, for example, to a better understanding of how fingertips receive and route information to the brain. This could in turn aid in the treatment of conditions like Carpal Tunnel Syndrome. An even further understanding of how the hand works could help in the development of robotic hands or software which will allow us to sense objects generated by a computer.

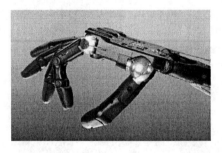

Figure 1-7 Robotic hand

Applied science is the science of applying the knowledge gained from one or more natural scientific fields to practical problems. The outcome of applied research can also be used to develop technology. One problem with extreme super-sonic flight (flying much faster than the speed of sound) is the heat generated on the surface of the material moving through the air. Therefore, research and development of materials having thermal properties sufficient to withstand such high temperatures in flight would be an example of applied science. Furthermore, once the material is developed it may show many more potential uses other than for aircrafts and could improve the technology of other systems.

Figure 1-8
U.S. Air Force SR 71 Blackbird has titanium skin which can withstand heat up to 1500 °C

Technological science would then determine the possibilities of what, if any, other applications the material could be used.

Chapter 1 Exercises

1. Describe in detail the components of the scientific method.

_____
_____
_____
_____
_____
_____
_____

2. Which is **not** one of the physical sciences?

   a. physics   b. chemistry   c. zoology   d. geology

3. Explain the difference between basic research and applied research

_____
_____
_____
_____
_____
_____
_____

4. Who invented penicillin?

_____

5. Define matter.

_____
_____
_____
_____

6. A team of scientists are looking at the possibility of using the material developed for space shuttles heat shield for use in the auto industry. What type of research is this?

_____

8. Science directed toward an immediate or practical goal is called

_____

9. Science that has no immediate or practical goal is called

_____

Chapter 2

The Chemistry of Matter

As mentioned in Chapter 1, matter is anything that has mass and occupies space. In this chapter we will explore different types of matter and the types of changes they can undergo.

The following key questions will be addressed:

- What are the properties of matter?
- How is matter classified?
- What are elements and pure substances?
- What are compounds?
- What are mixtures?
- What are physical & chemical properties?
- What are physical & chemical changes?
- What are chemical reactions?
- How do we use the symbols of elements in chemical compounds and equations?
- What is quantitative vs. qualitative?

Matter as we know it can be classified into three categories: macroscopic, microscopic and submicroscopic. **Macroscopic** is a form of matter large enough to be seen with the naked eye, whereas **microscopic** matter requires the aid of a microscope or similar instrument to be viewed. Viruses, bacteria and red and white blood cells are examples of microscopic matter. **Submicroscopic** matter is too small to be seen even with the most powerful microscopes. This type of matter is used to describe the material found in the make up of atoms and elements.

Mass is defined as the quantity of matter in an object. You are probably more familiar with the term **weight**, the force that results from the attraction between matter and the earth.

Figure 2-1
Although an astronaut is weightless in orbit his mass is the same as it was on earth's surface.

Therefore the weight of an object will vary from one place to another but the mass will remain constant. Although you would weigh six times less on the moon than you do here on earth your mass would be the same on both surfaces.

Figure 2-2
Laboratory balance

The mass of an object can be measured using a balance (figure 2-1) by comparing it to another object with a known or standard mass (figure 2-2)[2]. Since the standard and unknown are both subjected to the same gravitational force, the balance measures the mass and not the weight of the object.

Figure 2-3.
The standard Kilogram, a platinum-iridium cylinder kept at Sèvres, France, near Paris.

Anything that is observed or measured about a sample of matter is called a **property**. One classification of properties divides them into **intensive** and **extensive** groups. Intensive properties are those that are independent of the sample size such as concentration, density, and boiling and freezing points. Water will freeze at the same temperature regardless of its size or volume. However, an extensive property depends on the size or amount of the sample like mass, volume and energy. One gallon of gasoline will produce more heat than one teaspoon of gasoline when combusted.

Classification of Matter

All matter is either composed entirely of a single unique substance or a combination of different unique substances called elements. These **elements** are substances containing only one kind of **atom** (smallest unit found in elements). For instance, copper is one of the many natural occurring elements found on earth, and given a large enough quantity, it can be viewed macroscopically (i.e. copper penny).

Figure 2-4
1943 copper penny, in the 1970 the U.S. started copper plating zinc in the production of pennies.

However, since it is an element it must therefore contain only one kind of atom; in this case it is the copper atom which alone is a submicroscopic substance that can not be decomposed into simpler substances by normal chemical means. Furthermore, elements are also classified as

---

[2] kilogram, abbr. kg, fundamental unit of mass in the metric system, defined as the mass of the International Prototype Kilogram, a platinum-iridium cylinder kept at Sèvres, France, near Paris.

pure substances, which are defined as matter with fixed composition at the submicroscopic level. All other forms of matter are called **compounds** and can be decomposed into simpler substances by normal chemical means. For instance, the compound water (H₂O) can be decomposed via electrolysis into hydrogen and oxygen, which are both elements.

$$2H_2O_{(l)} \rightarrow 2H_{2(g)} + O_{2(g)}$$

Equation 2.1

Conversely the elements hydrogen and oxygen can combine to form water.

$$2H_{2(g)} + O_{2(g)} \rightarrow 2H_2O_{(l)}$$

Equation 2.2

Once elements combine to form a chemical compound their individual characteristics are replaced by those of the compound.

Figure 2-5. Hydrogen and oxygen are used as fuels to propel rockets into outer space.

For instance, hydrogen and oxygen uncombined, are very explosive (see Figure 2-5) whereas the combined form is not only inflammable but is used to extinguish fires.

Mixtures are also classified as either **heterogeneous** or **homogeneous**. A heterogeneous mixture is one in which different substances can easily be seen within the mixture. The mixture of oil and vinegar is heterogeneous, see figure 2-6.

Compounds, like elements can be pure substances or mixtures of different compounds which could be separated by chemical means into separate pure substances.

Figure 2-6.
Heterogeneous mixtures of oil and vinegar

Figure 2-7.
Window cleaner is a homogeneous mixture or solution.

Homogeneous mixtures have a uniform composition throughout the sample. For example, window cleaners are mixtures of several different compounds but have a uniform appearance, see Figure 2-7. Window cleaner is also an example of a homogeneous **solution.**

However, not all solutions are liquids; many alloys, mixtures of different metals, are solid solutions. A mixture of copper and tin, commonly called bronze (Figure 2-8) is a homogeneous mixture or a solid solution.

Figure 2-8.
Bronze statue of Auguste Rodin France 1840.

Furthermore, since all gases are homogeneous mixtures they are also solutions. Air is a solution composed mostly of nitrogen and oxygen. Since mixtures are not pure substances they can be decomposed into simpler substances or elements by normal chemical or physical means. For example a solution of salt water can be separated using the fact that water boils at 100 °C where salt boils at around 1600 °C. Heating a sample of salt water to 100 °C and allowing the water to evaporate will leave the salt behind.

Chemical and physical properties

Properties can be classified as either chemical or physical see Table 2-1. **Chemical properties** describe the tendency of a material to react and change into a different compound or set of new compounds. Flammability is a chemical property of gasoline. Once gasoline is combusted, it converts into energy and two new substances: carbon dioxide and water. **Physical properties** can be measured without changing the identity of the material. Melting point is a physical property of water, because when water changes from a solid to a liquid at 0 °C both the solid and the liquid are still *water*. This type of change (melting) is termed a **physical change** since it resulted in a change in only physical not chemical identity. When a change results in at least part of the substance being converted into a different kind matter it is called a **chemical change** (Table 2-2). For example the rusting of iron results in the formation of a new compound, iron oxide, therefore, rusting is a chemical change. See Figure 2-9.

Figure 2-9.
When iron is subjected to water and oxygen, rusting will occur.

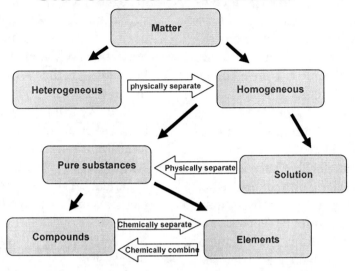

Figure 2-10.
Classification of matter

Chemical and physical properties

Chemical	Physical
Flammability	Boiling point
Inflammatories	Melting point
biodegradable	

Table 2-1 Chemical and physical properties.

Chemical and physical changes

Chemical	Physical
Cooking beef	Boiling
Rusting	Melting
Leaves changing colors	Freezing

Table 2-2 Chemical and physical changes

Chemical Reactions

The process in which one or more substances are converted to one or more different substances is called a **chemical reaction**. All chemical reactions have associated **reactants**, substances that undergo a change in a chemical reaction, and **products**, substances that are formed in a chemical reaction. For example when hydrogen and oxygen react they form water as a product (Equation 2-3).

$$2H_2 + O_2 \rightarrow 2H_2O$$
$$\text{Reactants} \qquad\qquad \text{Product}$$
Equation 2-3

Equation 2-3 shows that two molecules of hydrogen react with one molecule of oxygen forming two molecules of water. A **molecule** is defined as the smallest unit of a chemical compound that can exist independently and still maintain the characteristic properties of the compound. They are also defined as a combination of atoms joined together so tightly that they behave as a single particle. If all the atoms in the molecules are the same, the substance is an element. And if atoms of different elements combine to form a molecule, the substance is a compound. In Equation 2-3 the molecules hydrogen and oxygen are both elements and the molecule water is a molecular compound. Furthermore, when molecules are formed from two atoms they are called **diatomic** molecules. For instance the stable form of hydrogen, nitrogen, and the halogens (fluorine, chlorine, bromine and iodine) exist as diatomic molecules.

Although it is not shown in Equation 2-3, most chemical reactions also result in a change in energy. **Energy** is defined as the ability to do work. When gasoline is combusted, energy in the form of heat is released; this energy can then be used to do work, like making your car move. Energy is classified as either potential or kinetic. **Potential energy** is energy in storage by virtue of position or arrangement. The molecules that make up gasoline are a source of energy therefore, gasoline has potential energy and when it is burned the energy is released. **Kinetic energy** is the energy of objects in motion. As gasoline is burns its potential energy is converted to kinetic energy which can be used to perform work.

Symbols of Elements

The symbols of the elements are abbreviations for their names and consist of one or two letters with the first always capitalized and the second in lower case. For most of the elements the symbol is

an obvious abbreviation of the name. For example, the symbols for carbon, nitrogen, oxygen, fluorine, bromine and sulfur are C, N, O, F, Br and S respectively. However, some symbols are abbreviations of the elements ancient name like tungsten, the metal used as filaments in most standard light bulbs, is abbreviated as W, from the ancient name Wolfram. Other examples include, Na for sodium (natrium), Pb for lead (plumbum), Au for gold (aurum) and Sn for tin (stannum). Table 2-3 lists some common elements and their symbols.

Name	Symbol	Name	Symbol
Aluminum	Al	Fluorine	F
Barium	Ba	Gold	Au
Bromine	Br	Hydrogen	H
Calcium	Ca	Iodine	I
Carbon	C	Iron	Fe
Chlorine	Cl	Magnesium	Mg
Chromium	Cr	Mercury	Hg
Copper	Cu	Nickel	Ni
Nitrogen	N	Oxygen	O

Table 2-3. Names and symbols of some common elements.

There are seven nonmetallic elements that exist as two-atom entities; these are called **diatomic molecules**:

H_2- hydrogen F_2-   fluorine
O_2- oxygen Br_2-   bromine
N_2- nitrogen I_2-   iodine
Cl_2- chlorine

A **chemical formula** is a written combination of element symbols that represents the different atoms combined in a chemical compound. **Subscripts** are used in chemical formulas as numbers written below the line to show the numbers or ratios of atoms in a compound.

For example the chemical formula of water is H_2O, where it shows there are two atoms of hydrogen and one atom of oxygen in the molecular compound. Table 2-4 gives some common compounds and their chemical formulas.

Compound	Formula
Water	H_2O
Carbon dioxide	CO_2
Carbon monoxide	CO
Ammonia	NH_3
Methane	CH_4
Sulfur dioxide	SO_2
Butane	C_4H_{10}

Table 2-4 Common compounds and their chemical formulas.

The two types of chemical formulas most widely used, are molecular formulas and structural formulas. **Molecular formulas** are chemical formulas that represent molecules with atomic symbols and subscripts as shown in Table 2-4. **Structural formulas** are chemical formulas that show the connections between atoms in molecules with straight lines.

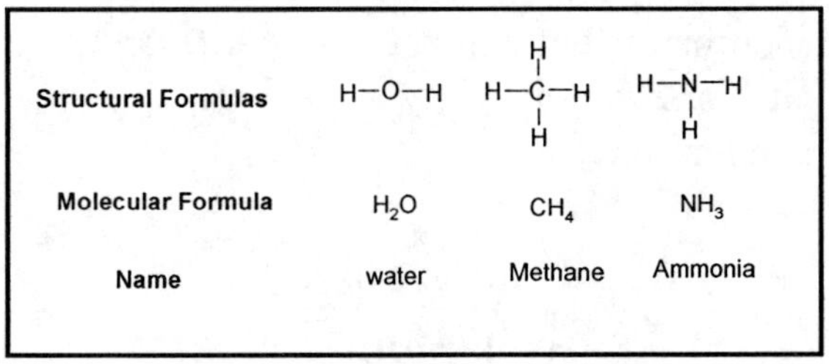

Figure 2-11.
Types of formulas for several common compounds.

Chemical Equations

Chemical changes can be completely described by equations. These **chemical equations** describe the identities and relative amounts of both reactants and products in a chemical reaction. For example hydrogen will react with chlorine producing hydrogen chloride.

$$H_2 + Cl_2 \rightarrow HCl \qquad \text{Equation 2-3}$$

The hydrogen (H_2) and chlorine (Cl_2) are the **reactants** and the hydrogen chloride (HCl) is a **product**. Notice that reactants are always to the left and products to the right of the reaction arrow. In general, two steps are required to write a proper chemical equation. First the identities of the reactant and products are determined and placed on the appropriate side of the reaction arrow. Second, because matter can neither be created nor destroyed (Chapter 3) the number and kinds of atoms must be the same on either side of the reaction arrow. When the two steps are complete the equation is said to be balanced. Inspection of Equation 2-3 reveals that it is not balanced. To balance the atoms we can add coefficients to precede the atoms of the reactants or products as needed. For example, adding the coefficient of 2 in front of the HCl molecule will balance both sides of the equation as shown in Equation 2-4.

$$H_2 + Cl_2 \rightarrow 2HCl \qquad \textbf{Equation 2-4}$$

Now, each side of the equation contains two hydrogen atoms and two chlorine atoms making the equation balanced. In Chapter 6, we will spend more time balancing equations and will discuss different relationships between the products and reactants.

13

Qualitative vs. Quantitative Measurements

As we proceed in the study of matter and the changes it undergoes, we need to decide on a method of observation. The two methods for measuring changes in matter are qualitative and quantitative. **Qualitative** methods are those that are not numerical but are used to identify the chemical species involved. When two species are allowed to react for the first time the potential products are unknown. The determination of the identity of the products would be a qualitative method. **Quantitative** measurements are those that are used to determine the amount of product formed or the amount of reactants used in a chemical reaction.

Determining the identity of the unknown is a qualitative procedure, whereas determining the amount of unknown formed is a quantitative procedure.

When quantitative measurements are taken it is important to use a standard set of units to describe quantities. The metric system, more commonly called the SI (system International) is on of the most often used in chemistry. The most common base or SI units and their abbreviations for different quantities are shown in Table 2-5.

Quantity	Unit	Abbreviation
Length	meter	m
Mass	Kilogram	kg
Volume	Liter	l
Temperature	Kelvin	K
Amount	Mole	mol

Table 2-5. Base or SI units for different quantities.

The SI creates units of different sizes by attaching a prefix to the base unit. For example the prefix Kilo- literally means for 1000 and can be used to change the unit from one size to another.

Unit Conversions

Changes are easily made among the SI units because the meanings of the prefixes provide conversion factors. Table 2-6 gives the prefixes used in the SI system.

Prefix	Abbreviation	Meaning
mega-	M	10^6
kilo-	k	10^3
deka-	d	10^1
centi-	c	10^{-2}
milli-	m	10^{-3}
micro-	μ	10^{-6}
nano-	n	10^{-9}

Table 2-6. Prefixes used with the SI system

A unit conversion factor is a fraction in which the numerator is a quantity that is equal or equivalent to the quantity in the denominator but expressed in a different unit. For example, sixty seconds is equal to one minute:

$$\left(\frac{60 \text{ seconds}}{1 \text{ minute}} \right)$$

This can be used to convert between units of seconds and minutes by multiplying by a value having the same unit as that found in the denominator of the unit conversion factor.

Example 2-1

Let's say you wanted to convert 29.5 minutes to units of seconds:

$$29.5 \; \text{minute} \; \times \left[\frac{60 \text{ seconds}}{1 \text{ minute}} \right]$$

$$= 1770 \text{ seconds}$$

Notice that the units of minutes cancel each other out since one is in the numerator and the other is in the denominator, leaving seconds as the only unit. This is what we wanted in the end. Let's now convert 1200 seconds to minutes:

$$1200 \; \text{seconds} \; \times \left[\frac{1 \text{ minute}}{60 \text{ seconds}} \right]$$

$$= 20 \text{ minutes}$$

Notice, this time we flipped the fraction so that the units of seconds would cancel and leave minutes.

Let's try another and convert 1.5 liters to milliliters. The unit conversion for liters and milliliters is

$$\left(\frac{1 \text{ liter}}{1000 \text{ milliliters}} \right)$$

So:

$$1.5 \; \text{liter} \; \times \left[\frac{1000 \text{ mL}}{1 \text{ liter}} \right]$$

$$= 1500 \text{ mL}$$

Notice once again we needed to flip the unit conversion to properly convert the units. Note that milliliter is abbreviated as ml using the base unit for liter (l) and the prefix (m) for milli-.

Uncertainty in measurements

All measured quantities have some uncertainty which often depends on the measuring devise used. For instance, if asked your current weight you might report the value from the last time you weighed yourself. However, the scale that you used might have some uncertainty. Have you ever weighed yourself on one scale and gotten a different result when weighed on another? It is important that when a measurement is made, the precision of the measurement must also be reported. This precision is sometimes given with the symbol ±. For example, the weight of a baseball might be reported as 145± 5.0 grams. This means that several measurements of the weight of a baseball fall between 140 and 150 grams. So, what is the difference between accuracy and precision? **Accuracy** is the term used to express the agreement of the measured value with the true value of the same quantity. For instance, if we wanted to determine the accuracy of a bathroom scale we might weigh an object with a known mass and see how close it reports. Repeated measurements are needed to determine the accuracy. For example let's say the object has a true mass of exactly 1.00 kg, and after 5 measurements were taken using the bathroom scale, the following data is collected:

Measurement	Result
1	1.05 kg
2	0.950 kg
3	1.03 kg
4	1.00 kg
5	0.970 kg
average	**1.00 kg**

The results show that the bathroom scale is very accurate. So, what is precision? Precision expresses the agreement among repeated measurements. For instance, let's say that we again measure the same 1.00 kg object on another bathroom scale and collect the following data:

Measurement	Result
1	0.910 kg
2	0.911 kg
3	0.909 kg
4	0.911 kg
5	0.0.909kg
average	**0.910 kg**

Notice that in this case the individual results are very close together (very precise), however, the average for all the results falls well below the actual value of 1.00 kg (not very accurate). So, this scale is not accurate but is very precise. Of course the goal is to have an instrument that reports values accurately and precisely.

Chapter 2 Exercises

1. List the four states of matter

_____
_____
_____
_____
_____

2. A pure substance which can be decomposed into two or more pure substances is a(n)

_____

3. Which of the following is considered a pure substance?

a. compound   b. homogeneous mixture   c. solution   d. heterogeneous mixture

4. Which is not a mixture?

a. pure water   b. mayonnaise   c. 14 kt Gold   d. ocean water

5. Matter that can be seen with the naked eye is called:

_____

6. Explain the difference between mass and weight.

_____

_____

_____

_____

_____

_____

_____

7. Give the definition of an extensive property?

_____

_____

_____

_____

_____

_____

_____

8. Which is a chemical property?

a. boiling point   b. odor   c. color   d. flammability

9. Define a pure substance.

_____

_____

_____

_____

_____

_____

_____

10. Which of the following is not a chemical change?

a. burning charcoal   b. rusting iron   c. melting ice   d. baking bread

11. Energy can be described as:

a. motion   b. heat   c. light   d. all of these

12. Define a heterogeneous solution?

_____
_____
_____
_____

13. What prefix is the largest?

a. mega   b. centi   c. micro   d. kilo

14. A person weighs 165 lbs. What is the weight in kilograms if 2.2 lbs. = 1 kg?

_____

15. Which prefix has the meaning 10^{-3}?

a. mega   b. nano   c. centi   d. milli

16. How many milligrams are there in 10 grams?

_____

17. The quantity 10^{-9} (one billionth) is designated by the prefix

_____

18. Convert 15 L of gasoline to gallons, given that 1.06 qt = 1 L ; 4 qts = 1 gal

_____

19. Explain the difference between accuracy and precision.

_____

_____

_____

_____

_____

_____

_____

20. Explain the difference between qualitative and quantitative.

_____

_____

_____

_____

_____

_____

_____

Chapter 3

A Closer Look at Atoms and the Periodic Table

In this chapter we will discuss the make up of atoms and how their individual properties are used to arrange them into similar groups. We will then define some of these groups and compare their properties to those of other groups. This of course will lead us into a discussion of the periodic table of the elements.

The following key questions will be addressed:

When, where and who developed the atomic theory?

What is the structure of the atom?

What are atomic and mass numbers?

What are isotopes?

What are atomic weights?

Where are electrons located in the atom?

How is the periodic table arranged?

What are periodic trends?

What are the main group elements and their properties?

Thousands of years ago Greek philosophers asked the question "is matter continuous or discontinuous"? Which is to say "can a sample of matter be divided into smaller and smaller pieces and still maintain the properties of the original sample?" If so, the matter is continuous, and if not it is discontinuous. Take a copper rod having properties such as electrical conductivity and a certain boiling and melting point. If the rod was initially a length of 1.0 meter and was cut in half would its properties stay the same? The answer, of course, is yes. However, is there a point when the sample is divided so much that the original properties are no longer the same? The answer to this question was proposed by **John Dalton** (1766- 1844) in his published paper on atomic theory. Dalton's atomic theory states:

John Dalton

1. Matter is composed of small indivisible particles called atoms. Atoms are the smallest units of an element that enter into a chemical combination.

2. An element is composed entirely of one type of atom. The properties of atoms of one element are different from those of any other element.

3. A compound contains atoms of two or more different elements. The relative number of atoms of

each element in a given compound is always the same.

4. Atoms do not change in chemical reactions. Chemical reactions involve changing the way in which the atoms are joined together.

Dalton's theory was used to explain many of the outcomes obtained during scientific experiments which resulted in formulating some important laws of chemistry. For instance, the chemical analysis of water finds a ratio of 8.0 grams of oxygen for every 1.0 gram of hydrogen. This is true no matter where the sample of water is taken. This observation is now known as the **law of constant composition**, that all samples of a pure substance contain the same elements in the same proportions by mass. This falls from Dalton's third postulate that states…. *"A compound contains atoms of two or more different elements. The relative number of atoms of each element in a given compound is always the same"*….

It is true that all samples of a specific compound contain the same elements in the same proportions, but sometimes more than one compound can be formed from the same elements. For example carbon and oxygen form carbon monoxide (CO) and can also form carbon dioxide (CO_2). Again, Dalton's theory was used to explain this important relationship that is called the **law of multiple proportions** which states that the masses of one element will always combine with a fixed mass of the second element in whole number ratios. For example take CO and CO_2. In CO, 1.33 grams of oxygen combines with 1.00 gram of car-

bon, whereas in CO_2, 2.66 grams of oxygen combines with 1.00 gram of carbon. Notice the the ratio 2.66/1.33 = 2; (a small whole number ratio). Dalton's fourth postulate helped explain the **law of conservation of mass or matter,** that there is no detectable loss or gain in mass when a chemical reaction occurs. As we saw in Chapter 2, chemical equations that describe chemical reactions are balanced when the same number and kind of atoms are present on both sides of the reaction arrow. If the number and kinds of atoms do not change in the course of a chemical reaction the mass cannot change.

Structure of the Atom

Although Dalton's theory describes atoms as being indivisible, it turns out that atoms are actually composed of three types of particles, called subatomic particles: electrons, protons and neutrons. This fact does not take away from Dalton's theory that dividing an atom into these individual particles will disrupt the properties to such an extent that will make it unrepresentative of the original atom. The properties of protons, electrons and neutrons are given in Table 3-1.

Particle	Mass (kg)	Relative mass	Relative charge
Electron	9.1×10^{-31}	0	-1
Proton	1.7×10^{-27}	1	+1
Neutron	1.7×10^{-27}	1	0

Table 3-1
　　Properties of protons, electrons and neutrons

Notice that the charge of the electron and proton are equal but opposite. Since

22

atoms are neutral species they must contain an equal number of electrons and protons. The mass of protons and neutrons are nearly the same[3], but the mass of the electron is roughly 1800 times smaller then either a proton or a neutron. Therefore, nearly all of the mass of an atom comes from the protons and neutrons.

Experiments by Ernest Rutherford (1871-1937) determined that the atom is mostly empty space, however, over 99% of the mass (contained in the protons and neutrons) is found in a very small region at the center of the atom called the **nucleus**. The diameter of the atom is about 100,000 times larger than the diameter of the nucleus, see Figure 3-2.

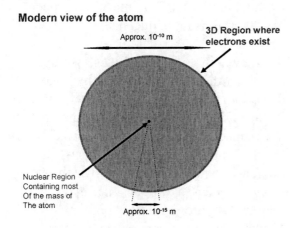

Figure 3-2
Modern view of the atom

Rutherford's experiment used beam of small subatomic particles (those smaller than the electron, proton, or neutron) directed through a gold foil where he showed that most of these particles passed right through without touching the foil, see Figure 3-3. This led to the modern view of the atom.

---

[3] The actual mass of a proton is 1.673×10^{-27} kg and that of the neutron is 1.675×10^{-27} kg.

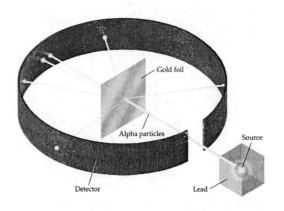

Figure 3-3
Rutherford's gold foil experiment

So, it is now understood that the protons and neutrons are found in the nucleus of the atom and the electrons are located in a three dimensional area around, at a relatively large distance away from the nucleus. This is called the **nuclear model of the atom.**

Atomic Number, Mass Number and Isotopes

The nuclear model allows us to describe the differences between elements. Different elements have different numbers of protons in the nucleus. It is actually the number of protons that determines the identity of the atom. This number of protons in an atom is called the **atomic number** and is represented by the letter Z. The **mass number** is defined as the number of protons plus the number of neutrons in the nucleus of an atom and is represented by the letter A. **Isotopes** are different atoms of the same element that contain different numbers of neutrons (same Z, different A). For example the hydrogen atom occurs naturally as three different isotopes. In each, the nucleus contains one proton; however, each isotope contains a different number of neutrons. For hydrogen the different isotopes are also named differently, see Figure 3-4.

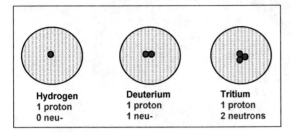

Figure 3-4 Three isotopes having atomic number Z=1: hydrogen, deuterium and tritium

All the naturally occurring elements exist as isotopes. Titanium for example has five different isotopes. In order to distinguish between these isotopes, a shorthand notation using the symbol of the element and the atomic and mass numbers are used having the form:

where X is the element's one or two letter symbol, A is its mass number (number of protons plus neutrons) and Z is the atomic number (number of protons). For the three isotopes of hydrogen the notation would be:

$$^1_1H \quad ^2_1H \quad ^3_1H$$

Hydrogen   Deuterium   Tritium

Since the number of protons determines the identity of the atom which has a unique symbol, the atomic number is sometimes omitted from the symbol of the isotope:

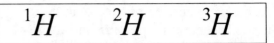

If the atomic number is omitted it can be easily determined by finding the symbol on the periodic table where the atomic number is found just above the symbol.

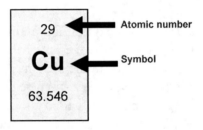

In the example above the atomic number for copper (Cu) is 29, meaning copper has 29 protons in its nucleus.

As shown in Figure 3-4 these isotopes are given the common names hydrogen, deuterium, and tritium. However, most isotopes do not have unique names but are named as the element followed by the mass number (i.e. hydrogen-1, hydrogen-2, and hydrogen-3, which are pronounced as hydrogen one, hydrogen two and hydrogen three respectively). Furthermore one can easily determine the number of neutrons in a nucleus from the name of the isotope. This is done by first determining the atomic number (number of protons) and subtracting it from the mass number (number of protons plus neutrons) which gives the number of neutrons in the isotope. For example, the carbon-14 (carbon fourteen) isotope has 8 neutrons in its nucleus. An inspection of the periodic table shows that the carbon (C) atom has an atomic number of 6. Subtracting this from the mass number of 14 gives 8, which is the number of neutrons.

Ions and Their Symbols

As was stated earlier, the numbers of protons and electrons are the same in all neutral atoms. But what happens if we remove either a proton or an electron from an atom? First, removing a proton will change the identity of the atom; for instance removing a proton from a carbon atom will change the atom into the element next to it called boron. This is much easier said than done because it requires an extreme amount of energy in order to remove a proton from a nucleus. However, it is much easier to remove an electron from an atom, and in doing so the atom retains its identity as well as its symbol, but, it is no longer a neutral species. In this case the atom will have a net charge of +1 because the protons out-number the electrons by one. Remember each proton has a charge of +1, and each electron has a charge of -1. Likewise, the addition of a proton to an atom is extremely difficult, and for this discussion we will assume it impossible. However, the addition of an electron is a much simpler task, and results in a net charge on the atom of -1, because now *the electrons out number the protons by one*. In chemical reactions, when atoms either gain or lose electrons, the new charged species are called **ions**. If atoms gain electrons, becoming negatively charged, they are called **anions**, whereas if they lose electrons, becoming positively charged, they are called **cations**. Symbols for ions are similar to those of isotopes because the magnitude of the charge is indicated as a superscript on the top right corner of the symbol. For example, when the calcium-41 isotope losses two electrons it acquires a net charge of +2 (the number of protons exceeds the number of electrons by two) and is given the following symbol:

$$^{41}\text{Ca}^{+2}$$

Example 3-1

Determine the atomic number, mass number, and number of protons, neutrons and electrons for the following species:

$$^{30}\text{Si}^{+4}$$

Answer:

The atomic number can be determined from the symbol by referring to the periodic table; Si with an atomic number of 14 is found directly below carbon (C). This is also equal to the number of protons. The mass number is equal to 30 as given by the superscript at the symbols top left corner. The number of neutrons can be found by taking the mass number and subtracting the number of protons: $30 - 14 = 16$. As for the number of electrons, in the neutral atom there are 14 (the same as the number of protons) however, since the charge (superscripts top right corner) is +4, meaning the protons outnumbers the electrons by 4. And since we cannot change the number of protons without changing the identity of the species, 4 electrons must have been removed from the neutral species; therefore there are 10 electrons found in this atom.

Atomic number =	14
Mass number =	30
Number of protons =	14
Number of neutrons =	16
Number of electrons =	10

Example 3-2

a)   Write the symbol for a species with 8 protons, 9 neutrons, and 10 electrons.

b)   Write the symbol for a species with 20 protons, 20 neutrons and 18 electrons.

Answer:

a)   The eight protons define the atomic number as 8, so the atom is oxygen (O). The sum of the number of protons and neutrons is equal to the mass number, $8 + 9 = 17$. There are two more electrons than protons so the charge is -2:

$$^{17}_{8}O^{-2} \text{ or } ^{17}O^{-2}$$

b)   The species has an atomic number of 20, which lets us know it is calcium (Ca). The mass number is 40 (number of protons + neutrons). Since there are two more protons than electrons the charge is +2:

$$^{40}_{20}Ca^{+2} \text{ or } ^{40}Ca^{+2}$$

Atomic Mass Units, Isotopic Distribution, and Atomic Weights

The **atomic mass unit** or (μ) has been established to compare the mass of any atom to that of the carbon-12 (^{12}C) isotope. The mass of the carbon-12 isotope is defined as exactly 12μ. Note that the atomic mass unit is the same as the mass number. This is no coincidence because one μ has been set numerically equal to the mass of a proton or neutron, and since there are 6 protons and 6 neutrons in the carbon-12 atom, therefore ^{12}C has an atomic mass of 12μ. So, using this scale ^{27}Al has an atomic mass of 27μ.

As was stated earlier, most naturally occurring elements exists as a mixture of isotopes. And for each element the ratios of these isotopes are constant in any sample of an element. For instance, it has been determined that the percentages of hydrogen's three isotopes in a naturally occurring sample are 99.6% ^{1}H, 0.25% ^{2}H, and 0.15% ^{3}H. (The sum of the individual percentages has to equal 100; 99.6% + 0.25% + 0.15% = 100%). This isotopic distribution will be the same in any sample of hydrogen obtained in nature. Another way of looking at this is to say that 99.6% of the hydrogen atoms in any sample of hydrogen has an atomic mass of 1μ, 0.25% has an atomic mass of 2μ and 0.15% has an atomic mass of 3μ. This distribution of isotopes in a sample can be used to calculate an average mass or atomic weight of a sample. An **atomic weight** is the weighted average mass in atomic mass units the isotopes of an element. It is calculated by multiplying the fraction that each isotope exists in a sample by its atomic mass unit and then adding each of the results together. For example, the atomic weight of hydrogen is:

$$(0.996 \times 1) + (0.0025 \times 2) + (0.0015 \times 3)$$

$$= 1.0055\mu \text{ or rounding gives } \mathbf{1.01\mu}$$

Example 3-3

Chlorine has two stable isotopes: ^{35}Cl with a natural abundance of 77.5% and ^{37}Cl with a natural abundance 22.5%. Calculate the atomic weight of Chlorine.

Answer:
The atomic weight is the weighted average of the two isotopes:

So, the atomic weight of Cl

$= (35 \times 0.775) + (37 \times 0.225) = 35.45\mu$

The atomic weight is always numerically equal to the number found under the symbol of each element on the periodic table.

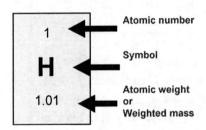

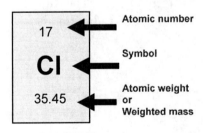

Location of Electrons in Atoms

Now, hopefully, a clearer understanding of the atom is starting to emerge. This will lead to a greater appreciation of the periodic table. However, before we introduce the periodic table and all of its glory, we must first look into the nature of the electron. As mentioned in Chapter 1, chemistry is the study of matter and the changes it undergoes. Many of these changes result in a chemical reaction involving the atoms of different samples of matter. A more detailed discussion of chemical reactions is found in Chapter 6. For now it is important to know that most chemical reactions involve the interactions of electrons in one sample of matter with electrons in a different sample. So, in order to understand how electrons interact, we must first become familiar with electrons and their locations in individual atoms.

First, let's explore the behavior of light. In the 19th century physicists experimented with and determined that light had characteristics similar to waves. Like the waves produced when a rock is thrown into a lake, light has associated up and down motions as it moves in space as shown in Figure 3-5.

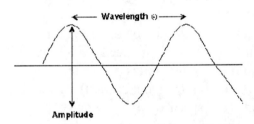

Figure 3-5 Typical wave pattern of light

These **waves** repeat at regular intervals of time and distance. Any wave can be described by its amplitude, wavelength and frequency as shown in Figure 3-4. The **amplitude** is the maximum height of the wave, and the **wavelength**, (λ, "lambda") is the distance between one peak and the next. Most wavelengths are measured in meters. The **frequency** (ν, "nu") is the number of waves that pass a particular point in a given time period; frequency is measured in hertz (Hz) having units of 1/s (one over seconds)

Light waves are a form of **electromagnetic radiation** in that they consist of oscillating electric and magnetic fields perpendicular to one another, as shown in Figure 3-6.

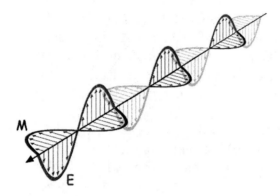

Figure 3-6 light or electromagnetic radiation consists of perpendicular magnetic and electric fields

Light or electromagnetic radiation comes in a variety of wavelengths, amplitudes and frequencies. The one thing they all have in common is their speed, which in a vacuum is 186,000 miles per second or 3.0×10^8 m/s (meters per second). This constant (the speed of light; denoted as "c") is the same for all forms of electromagnetic radiation. It also turns out that the product of the wavelength times the frequency is always equal to the speed of light, 3.0×10^8 m/s. This relationship gives rise to the following equation for frequency and wavelength.

$$c = \lambda \nu$$

Speed of light = wavelength (meters) times frequency (hertz)

Example 3-4

Calculate the frequency for a wave having length of 3.21 meters:

Answer:

Use the relationship $c=\lambda\nu$ and rearrange for frequency (ν)

$$\nu = \frac{c}{\lambda} = \frac{3.0 \times 10^8}{3.21} = 93457943 \text{ Hz}$$

or 93.5×10^6 Hz

or 93.5 MHz (megahertz)

Note: this is the frequency for a local radio station. And it is also a radio wave.

In the previous example the frequency of the electromagnetic radiation corresponds to a radio wave. Electromagnetic radiation can vary from very short wavelengths (high frequency) to very large wavelengths (low frequency). The full range of electromagnetic radiation is shown in Figure 3-7.

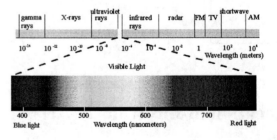

Figure 3-7 Range of electromagnetic radiation. Note that the visible light occupies a very small region of the spectrum.

Now, how does all of this apply to electrons? Experiments have proven that electrons behave like and have very similar characteristics to light waves. So, as light waves can interact with other light waves, they can also interact with electrons. The outcome of this interaction can have a positive or negative effect. A positive outcome will result in the combination of the individual waves creating a more intense light. A negative outcome will result in the waves canceling each other and diminishing in intensities. This phenomenon can be viewed easily when light from a flashlight is pointed to a wall, creating rings of light and dark; where the light rings are a result of **constructive interference** (positive effect), the dark rings are the result of **destructive interference** (negative effect). Furthermore, white light (the absence of color), consists of all the frequencies of light in the visible region, is a result of destructive interference, and becomes evident when passed through a prism and splits into separate colors. These colors are due to the different frequencies which are related to their energies. Examination of the electromagnetic spectrum shows that gamma rays, which are very high energy waves and can cause serious damage to human tissue, are located to the far left of the spectrum. On the other hand, radio waves, which are low energy waves, are located to the far right. Therefore, as we move from left to right along the spectrum the light waves have less energy. We can infer that blue light, found to the left of the visible region is higher in energy than red light which is farther to the right side of the spectrum. Now let's see what happens when light interacts with electrons. When light comes in contact with an electron, the electron will absorb the energy from the light wave and moves to a higher energy state or **excited state**. These exited electrons can either stay in their current state or release the energy obtained and move to their lower original state, called the **ground state**. When this happens the emitted light from the electron can be detected and its energy determined. The electrons in atoms will only absorb energy of a certain wavelength when moving to a particular excited state. These excited states, or higher locations of the electrons, are separated by discrete energy levels (n) or sometimes called **principle shells,** as shown in Figure 3-7.

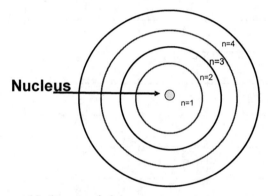

n= 1 is the ground state
n= 2 first excited state, etc.

Figure 3-8 Excited states for electrons around the nucleus of an atom.

The model of the atom as shown in Figure 3-8 was proposed by **Niels Bohr** in 1911 when he showed that electrons in atoms were limited to these energy levels or shells. Bohr also showed that there is a maximum number of electrons that can occupy any given energy level as determined by:

$$\text{Max}_{e^-} = 2(n)^2$$

(Where, n is the energy levels found in the Bohr atom, see figure 3-7)

In the principle energy shell n=1 the maximum number of electrons that can occupy this level is:

Max $_{e-}$ = $2(n^2)$ = $2(1^2)$ = 2 electrons

For n=2

Max $_{e-}$ = $2(n^2)$ = $2(2^2)$ = 8 electrons

Table 3-2 gives the maximum number of electrons for each energy level.

Level=n	$2n^2$	Max. # of e-
1	$2(1^2)$	2
2	$2(2^2)$	8
3	$2(3^2)$	18
4	$2(4^2)$	32
5	$2(5^2)$	50
6	$2(6^2)$	72
7	$2(7^2)$	98

Table 3-2 Maximum number of electron for the principle energy shells n=1 to n=7

Now, we can talk about the locations of electrons in atoms found on the periodic table. Let's look at some examples of neutral atoms. Hydrogen has one electron and in the ground state will occupy the n=1 principle shell of the Bohr model. The two electrons found in helium (He) can also occupy the n=1 principle shell. However, lithium (Li) has 3 electrons, two of which will occupy the n=1 level the other will occupy the n=2 level. This allows us to describe the location of electrons about an atom using the Bohr configuration of electrons. For example

the electrons in the carbon atom can be described as:

2-4

Carbon has six electrons in its neutral state, two of which are in the n=1 level and the other four are in the n=2 level.

Atomic Orbitals

Electrons occupying different energy shells (n levels) are located in atomic orbitals. Where an **atomic orbital** is a region of 3-dimensional space where electrons exist around the nucleus. Each principle energy shell consists of a certain number and type of orbitals grouped into **subshells**. The number of subshells within each principle energy level is equal in number to the value of n. For instance, for the n = 1 principle shell there is only one subshell available and for the n = 3 principle shell there are three subshells available and numbered 1, 2, and 3 respectively. In other words the number of subshells available for each principle shell (n) is equal to 1, 2, 3, ...n. Each subshell contains a certain number of atomic orbitals. The number of atomic orbitals within each subshell is equal to 2 times the subshell number (Ss#) minus one:

Number of orbitals = (2 x Ss#) - 1

For example, principle shell (n = 1) contains only one subshell and is numbered 1. Therefore the number of atomic orbitals in this subshell is:

(2 x 1) – 1 = 1

The n = 2 principle shell has 2 subshells numbered 1 and 2. The number 1 subshell has only on atomic orbital, like in

30

the previous example, however, for the number of atomic orbitals in the numbered 2 subshell is:

$$(2 \times 2) - 1 = 3$$

Furthermore, each subshell has been given a letter designation based on its subshell number. Subshells numbered 1, 2, 3, and 4 have been assigned the letters s, p, d, and f respectively. In order to specify which principle shell these subshells are associated the principle shell number precedes the letter of the subshell. For instance, a *d* (Ss# = 3) subshell in the n = 3 principle shell would be written as 3d.

Example 3-5

How many atomic orbitals are in the 3d subshell?

Answer

A d subshell correlates to a subshell number (Ss#) = 3. The number of orbitals would therefore be:

$$(2 \times 3) - 1 = 5$$

So, the 3d subshell contains 5 atomic orbitals.

Table 3-3 shows the relationship for subshells, letter designations and number of atomic orbitals per subshell.

n level	Subshells	Subshell letter	orbitals per subshell
1	1	s	1
2	1, 2	s, p	1, 3
3	1, 2, 3	s, p, d	1, 3, 5
4	1, 2, 3, 4	s, p, d, f	1, 3, 5, 7

Table 3-3 relationships for atomic orbitals and subshells

The maximum number of electrons that can occupy any given atomic orbital is two. An *s* subshell can therefore hold only two electrons, where a *p* subshell can contain a maximum of six electrons. Table 3-4 gives the maximum number of electrons for each orbital type.

Subshell	# of orbitals	Maximum # of electrons
s	1	2
p	3	6
d	5	10
f	7	14

Table 3-4 maximum numbers of electrons that can occupy a given atomic orbital

Energy of Atomic Orbitals

The relative energy of atomic orbitals is based on the principle energy level (*n*). However, as electrons are added around an atom the energy of the atomic orbitals tend to shift due to electron repulsion. This shift is exaggerated at the *n* = 3 principle energy level where the energy of the 3d subshell rises just above the energy of the 4s subshell.

31

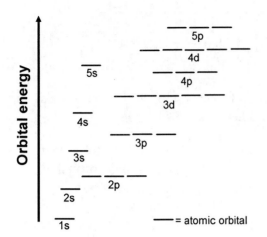

Figure 3-9 orbital energies

This shift is also seen for the $n = 4, 5$ and 6 principle energy levels. Figure 3-9 shows the relative energy of atomic orbitals when occupied by electrons.

Electron Configuration

Electrons are added to atoms one at a time to available orbitals with the lowest energy first. This process is called the **aufbau principle**, from the German word *aufbau* meaning building up. The distribution of these added electrons is termed the **electron configuration** for the atom. For example let's take an atom with six total electrons about the nucleus. The first two electrons will occupy the 1s atomic orbital, the next two will be placed in the 2s atomic orbital and the remaining 2 electrons will partially fill the 2p atomic orbital, see Figure 3-9.

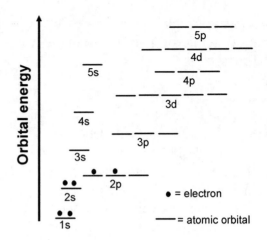

Figure 3-10 filling of atomic orbitals for an atom with six electrons.

Figure 3-10 Also shows that the two electrons in the 2p subshell are placed in separate orbitals first and only *paired* when the fourth electron is added to this subshell.

The electron configuration for this atom would therefore be written as:

$$1s^2 2s^2 2p^2$$

This shows that two electrons are located in the 1s orbital, two electrons in the 2s orbital and two electrons in the 2p orbital.

Example 3-6

What is the electron configuration for an atom with 23 electrons?

Answer
Remember the maximum number of electrons that can occupy an s-orbital is two, a p-orbital is six and a d-orbital is ten. Therefore the electron configuration is;

$1s^2 2s^2 2p^6 3s^2 3p^6 4s^2 3d^3$

Note; the sum of the superscripts has to equal the total number of electrons.

Valence Electrons

Orbitals of the outermost or highest energy level and partially filled subshells of lower energy are called **valence orbitals**. Electrons found in these valence orbitals are called **valence electrons**. Valance electrons are the ones most involved in chemical reactions. In the example above there are five valence electrons located in the n = 4 ($4s^2$, highest occupied energy level) and n = 3 ($3d^3$, partially filled lower energy level).

Example 3-7

How many valence electrons are in an atom with 13 electrons?

Answer

Thirteen electrons would have an electron configuration of;

$$1s^2 2s^2 2p^6 3s^2 3p^1$$

The highest energy level is 3, so electrons found in n = 3 subshells are valence electrons. There are two in the 3s subshell and one in the 3p subshell. Since there are no partially filled subshells of lower energy the number of valence electrons is three (3).

The Periodic Table

Now that you are familiar with the structure of atoms, and their ability to combine to form elements and compounds, we will now discuss how the properties of atoms are related to each other. Most of the known elements were isolated by chemists in the middle 19th century. As scientists examined the chemical and physical properties of these elements, they found that many were quite similar. For example, lithium, sodium, and potassium were found to have very similar reactivities, but were very different from fluorine, chlorine, and bromine which also showed very similar chemical properties.

In the early 1900's Russian scientist Dimitri Mendeleev proposed arranging the elements into rows and columns based on their similar physical and chemical properties.

Dimitri Mendeleev (1834-1907)

Mendeleev arranged the lighter elements at the top of a column and the heavier ones at the bottom. These columns were then connected into rows which showed regular changes of chemical properties across from left to right. Each row of the table is called a **period** and is numbered from one to seven. The columns are called groups and numbered across the top of the table. The modern table uses Roman numerals and letters to label the columns. Figure 3-11 shows the modern periodic table. The elements in the table are further divided into metals and nonmetals. A **metal** is generally defined as a substance that is malleable, has luster, and is a good conductor of electricity.

Periodic Table of The Elements

Atomic Number → 1
H ← symbol
Atomic weight → 1.01

IA																	VIIIA
1 **H** 1.01	IIA											IIIA	IVA	VA	VIA	VIIA	2 **He** 4.026
3 **Li** 6.94	4 **Be** 9.01		☐ Metals ☐ Metalloids ☐ Nonmetals									5 **B** 10.81	6 **C** 12.01	7 **N** 14.01	8 **O** 15.99	9 **F** 18.98	10 **Ne** 20.18
11 **Na** 22.99	12 **Mg** 24.31	IIIB	IVB	VB	VIB	VIIB	—— VII ——		IIB	IB		13 **Al** 26.98	14 **Si** 28.09	15 **P** 30.97	16 **S** 32.07	17 **Cl** 35.45	18 **Ar** 39.95
19 **K** 39.09	20 **Ca** 40.08	21 **Sc** 44.96	22 **Ti** 47.88	23 **V** 50.94	24 **Cr** 51.99	25 **Mn** 54.94	26 **Fe** 55.85	27 **Co** 58.93	28 **Ni** 58.69	29 **Cu** 63.54	30 **Zn** 65.39	31 **Ga** 69.72	32 **Ge** 72.61	33 **As** 74.92	34 **Se** 78.96	35 **Br** 79.90	36 **Kr** 83.80
37 **Rb** 85.47	38 **Sr** 87.62	39 **Y** 88.91	40 **Zr** 91.22	41 **Nb** 92.91	42 **Mo** 95.94	43 **Tc** (98)	44 **Ru** 101.1	45 **Rh** 102.9	46 **Pd** 106.42	47 **Ag** 107.8	48 **Cd** 112.4	49 **In** 114.8	50 **Sn** 118.7	51 **Sb** 121.7	52 **Te** 127.6	53 **I** 126.9	54 **Xe** 131.3
55 **Cs** 132.9	56 **Ba** 137.3	57 **La** 138.9	72 **Hf** 178.5	73 **Ta** 180.9	74 **W** 183.9	75 **Re** 186.2	76 **Os** 190.2	77 **Ir** 192.2	78 **Pt** 195.1	79 **Au** 196.9	80 **Hg** 200.6	81 **Tl** 204.4	82 **Pb** 207.2	83 **Bi** 208.9	84 **Po** (209)	85 **At** (210)	86 **Rn** (222)
87 **Fr** (223)	88 **Ra** (226)	89 **Ac** (227)	104 **Rf** (261)	105 **Ha** (262)	106 **Sg** (263)	107 **Ns** (262)	108 **Hs** (265)	109 **Mt** (266)									

58 **Ce** 140.1	59 **Pr** 140.9	60 **Nd** 144.2	61 **Pm** (145)	62 **Sm** 150.4	63 **Eu** 152.0	64 **Gd** 157.3	65 **Tb** 158.9	66 **Dy** 162.5	67 **Ho** 164.9	68 **Er** 167.3	69 **Tm** 168.9	70 **Yb** 173.0	71 **Lu** 175.0
90 **Th** 232.0	91 **Pa** 231.0	92 **U** 238.0	93 **Np** (237)	94 **Pu** (244)	95 **Am** (243)	96 **Cm** (247)	97 **Bk** (247)	98 **Cf** (251)	99 **Es** (252)	100 **Fm** (257)	101 **Md** (258)	102 **No** (259)	103 **Lr** (260)

Figure 3-11 periodic table of the elements

Metals are primarily found in the middle and the left hand side of the periodic table, while nonmetals, elements that typically do not conduct electricity, are found in the top right corner of the table. The table is further divided into **metalloids**, which fall along the actual line separating the metals from the nonmetals and share certain properties of both. One important property of metalloids is that they are **semiconductors** or weak conductors of electricity, which allows them to be used extensively in solid-state electronics. Another class of elements of the periodic table is the **transition metals**, a group of elements showing a regular decrease in metallic behavior; they are found in the center of the table in groups IIIB to IIB. The elements in groups IA through VIIIA, are called the **representative elements** or main **group elements**. The **inner transition** metals are found at the bottom of the table and are named the **lanthanides** (cerium, Ce through lutetium, Lu) and the **actinides** (thorium, Th through lawrencium, Lr). These metals do not belong to any of the groups of the main table and most of the actinides are not found in nature but are made in a laboratory

Important Groups of the Representative Elements

Elements of group IA are called the **alkali metals**. These metals are very reactive, especially with water, and must be handled carefully. Their reactivity's tend to increase down the group. They are soft and malleable metals having relatively low melting points compared with other metals. Because of their reactivities they always occur as compounds. Sodium (Na) and potassium (K) are very abundant in the earth's crusts and are generally in the form of salts, i.e. sodium and potassium chloride, NaCl and KCl respectively. **The alkaline earth metals** are found in group IIA and are slightly less reactive than the alkali metals but are still very reactive. Magnesium (Mg) and calcium (Ca) are also very abundant in the earth's crust and are important components of coral, bones, and seashells. The **halogens** (group VIIA) are among the most reactive nonmetals with reactivity's decreasing down the group. Furthermore, as you move down this group the halogens exist as gaseous (fluorine and chlorine) as a liquid (bromine) and finally as a solid (iodine) at room temperature. Chlorine is the most abundant of the halogens and is a major component in table salt, (NaCl) and seawater. The elements in group VIIIA are called the **noble gases**. At one time these elements were called "inert gases" because of their seemingly non reactive behavior. However, in recent years some of these elements (xenon, Xe and krypton, Kr) have shown to form compounds, so the term inert was replaced with noble.

Periodic Trends of the Elements

The arrangement of the elements in the periodic table based on their similar chemical and physical properties has indeed led to a better understanding of the nature of chemical reactions. However, several questions still need to be answered. Like, why do groups of atoms like the alkali metals or the halogens have similar properties? Or, why do the metals and nonmetals have such different properties?

Valence Electrons and the Periodic Table

These questions were answered when G.N. Lewis studied the concept of valence electrons. Recall that valence electrons are those occupying valence orbitals. Lewis proposed that because noble gases (group VIIIA) have completely filled outer shells giving the same number of valence electrons for each noble gas in the group and results in their very low reactivity and therefore considerable stability. He further proposed that the reactivities of other elements were influenced by their numbers of valence electrons. We will see in Chapter 5 that most chemical combinations are a result of atoms achieving or trying to achieve a noble gas configuration.

Expanding on electron configuration of electrons and the modern form of the periodic table we can see a correlation exist between the arrangement of atoms and number of valence electrons. For example, the neutral hydrogen atom has one electron giving it an electron configuration of $1s^1$, lithium (three electrons) has an electron configuration of $1s^12s^1$, also with one electron in its valence orbitals, and sodium's configuration is $1s^22s^22p^63s^1$, again having one electron in its valence orbital. Stemming from the facts that all of these have one electron in their valence orbitals and that all

35

appear in the same group (IA), led to the conclusion that atoms with the same number of valence electrons share similar chemical and physical properties. The numbers of valence electrons are therefore reported on the periodic table at the top of each group as a Roman numeral. For instance, all atoms in group IA have one valence electron and all in group VIIA have seven valence electrons.

Example 3-8

a) How many valence electrons does the phosphorous (P) atom have?

b) Name the first atom in the group with three valence electrons.

Answer:

a) Phosphorous is found in group VA, therefore it has 5 valence electrons.

b) The group of atoms containing three valence electrons belongs to group IIIA, the first atom in this group is boron (B).

Other Properties of Atoms

As was stated earlier all atoms desire a noble gas electron configuration, which is accomplished by either chemical combination or by the loss or gain of electrons. The specific course taken to achieve "nobility" can be related to the elements number of valence electrons. For instance, sodium (Na) has one valence electron, so for it to achieve a noble gas configuration it will have to either lose one or gain seven electrons. It is more likely that sodium will lose the single electron and as a result obtain a plus one (+1) charge. Conversely chlorine (Cl) would have to either gain one or lose seven electrons to achieve a noble gas configuration.

Again, the easier way is to be altered by one electron, in this case gain it and acquire a negative one (-1) charge. It turns out that the ease at which these atoms gain or lose electrons is related to their sizes and consequently to their abilities to react.

Atomic Radii

As you move down the elements of group IA their atomic radii increase. This is because the larger elements have larger atomic numbers and more energy levels (shells) which contain electrons further and further away from the nucleus. However, the size of atomic radii tends to decrease across a period from left to right. This might seem confusing because as you move from left to right, more protons, neutrons and electrons are being added to the atoms which increase the atomic number. You might assume that the atom would grow larger, but these additions have very little effect on the size of the electron cloud (the 3-D area of electrons surrounding the nucleus of the atom). The reason for the decrease in size is that when more protons are added to the nucleus while electrons are added to a given energy level (n-level), the attractive forces that the protons have on the electrons increase thereby decreasing the distance the electrons are from the nucleus of the atom. Figure 3-12 shows the sizes of the atoms of the representative elements.

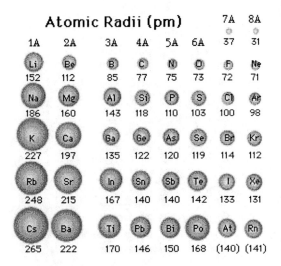

Figure 3-12 Atomic radii for the atoms of the representative elements.

Reactivity

Since reactivity is based on the atoms' ability to gain or lose electrons, it follows that the larger atoms would lose electrons easier than the smaller atoms. The valence electrons in larger atoms are farther away from the nucleus and are therefore less tightly held by the protons. This would suggest that cesium (Cs) the largest metal would be the most reactive, which is indeed the case. On the other hand, achieve a noble gas configuration is easier for smaller atoms because the protons would affect a greater pull on the incoming electron or electrons. Therefore, as you move from top to bottom, on the periodic table, the reactivities of the metallic groups increase, but decrease for the non-metallic groups.

Ionic Radii

It is now clear that the representative elements tend to form compounds as ions (by either gaining or losing electrons). The elements in groups IA, IIA, IIIA will lose one, two and three electrons and form ions having a charge of +1, +2 and +3 respectively. The elements of groups VA, VIA, and VIIA will gain three, two, and one electron and form ions with a -3, -2 and -1 charge respectively. Since we have attributed the atomic size to the ability of the protons in the nucleus to pull electrons closer, thereby reducing the size of the electron cloud, we can now account for the size of ions. As the number of electrons decreases about an atom while the number of protons remains constant, the size of the electron cloud is reduced; this reduces the size of the ion. In other words cations (positively charged ions) are smaller than their neutral atoms. On the other hand, when electrons are added around an atom the effect the protons have on them is reduced which allows the electron cloud to increase making the ion bigger; anions (negatively charged ions) are larger than their neutral atoms. Figure 3-13 shows the charges of the ions formed by the representative elements with their radii in picometers and angstroms in parentheses.

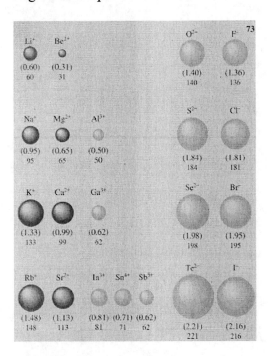

Figure 3-13 radii of ions of the some of the representative elements

Ionization Energies

As was stated earlier the formation of cations (positively charged ions) is from the loss of an electron or electrons. This process requires a certain amount of energy called the **ionization energy**. This is the amount of energy needed to remove an electron and increases from left to right and decreases from top to bottom on the periodic table as shown in Figure 3-12.

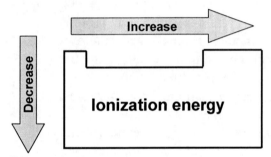

Figure 3-12 Ionization energy trend across the periodic table

Chapter 3 Exercises

1. In a neutral atom there is an equal number of

   a. protons and electrons   b. neutrons and protons   c. electrons and neutrons   d. protons, electrons, and neutrons

2. Rutherford's foil experiment provided evidence for what atomic feature?

   _____

3. Which sub-atomic particle has the incorrect charge indicated?

   a. proton - positive   b. electron - negative   c. neutron - neutral   d. nucleus - neutral

4. In the atom represented by the symbol ^{23}Na, there are how many protons, neutrons and electrons

   _____

   _____

   _____

5. What is the symbol for a neutral atom with 7 protons, 7 electrons, and 8 neutrons?

   _____

6. Which of the following is not included with a periodic table entry?

   a. atomic number   b. mass number   c. atomic weight   d. symbol

7. What isotope is used to define the atomic mass unit scale?

   _____

8. Which statement is true about the speed of photons and light waves?

a. red light is slower than blue light   b. each color travels at a different speed   c. all travel at the same speed   d. the longer the wavelength, the lower the speed

9. Which of the following has the longest wavelength?

a. radio waves   b. X-rays   c. visible light   d. ultraviolet

10. Which relationship is correct?

a. lower wavelength, small energy   b. lower frequency, lower wavelength   c. higher wavelength, higher frequency   d. higher frequency, larger energy

11. The third energy level can hold a maximum of how many electrons?

_____

12. Write the electron configuration for the following elements; a) B b) Cl c) Ti d) Br

_____

_____

_____

_____

_____

_____

_____

13. What is the number of valence electrons in phosphorus atom (P) having an electron configuration of $1s^2 2s^2 2p^6 3s^2 3p^3$?

_____

14. Which color of light carries the most energy per photon?

a. blue   b. green   c. orange   d. red

15. Which color of light carries the least energy per photon?

a. blue   b. green   c. orange   d. red

16. Who developed the atomic theory?

    _____

17. Carbon and oxygen always form in ratios of small whole numbers is an example of which law?

    _____

18. Which wavelength of visible light is closest to red?

    a. 400 nm   b. 500 nm   c. 700 nm   d. 300 nm

19. Which is not found in the nucleus of an atom?

    a. neutron   b. proton   c. electron   d. ion

20. Which of the following ground state configurations matches represents an atom with 3 valence electrons?
    a. 2-8-3   b. 2-1   c. 2-3-5   d. 2-8-8

21. What metal was used as the foil in Rutherford's famous scattering experiment?

    _____

22. The name given to the number of protons in an atom's nucleus is

    a. atomic number   b. family number   c. electron number   d. mass number

23. Two atoms which have the same atomic number but different mass numbers are called

    _____

24. What name is given to the sum of neutrons and protons in an atom's nucleus?

    _____

25. The modern periodic table is based on arranging elements in the order of their

_____

26. Which element is found in Group IIA - in Period 4?

a. magnesium, Mg   b. zinc, Zn   c. calcium, Ca   d. potassium, K

27. Write the symbol for a species with 7 protons, 8 neutrons, and 11 electrons?

_____

28. Which element is a noble gas?

a. Ni   b. Ne   c. Si   d. B

29. How many valence electrons are in boron, B

_____

30 Which of the following atoms is the smallest?

a. Ar   b. Mg   c. Cl   d. Si

31 What element (X) is a metal that forms +1 ions?

a. sodium (Na)   b. aluminum (Al)   c. calcium (Ca)   d. nitrogen (N)

32 Sodium has chemical properties most like

a. cesium, Cs   b. magnesium, Mg   c. chlorine, Cl   d. mercury, Hg

33 Which of the following is a transition element in the periodic table?

a. sodium b. sulfur c. boron d. chromium

34 Which of the following is an atom in Group IVA?

a. C b. Cu c. Cs d. Cl

35 List the alkaline earths metals?

_____

_____

_____

36 Which atom in the following series is the largest?

a. K b. Rb c. Cs d. Na

37 Which of the following has the highest ionization energy?

a. K b. Mg c. Si c. S

38 Which of the following reacts most violently with water?

a. Ne, neon b. Na, sodium c. Li, lithium d. K, potassium

39. Which of the following elements has five valence electrons?

a. Be b. F c. P d. Xe

44

Chapter 4

Nuclear Chemistry, the Chemistry of Neutrons and Protons

Most people attach a negative connotation to the word 'nuclear'. This is understandable since most things associated with this word are undesirable. Nuclear war, nuclear disaster, nuclear radiation, nuclear waste, and nuclear fallout, which can lead to a nuclear winter, are just a few stigmas to the term nuclear.

Figure 4-1
View of the vapor cloud produced from the atomic bomb

It wasn't until the splitting of the atom in the 1940's and the eventual making of the atomic bomb that these negative associations with the word nuclear came to be. However, even with all the negative stereotypes associated with this word there are more examples of how the technology of the 1940's has had a positive impact on our lives. For instance, nuclear power has given electricity to people who live in areas that would not have otherwise been able to generate it.

Figure 4-2
Nuclear power plant

Thousands upon thousands of lives have been saved from various forms of cancer using nuclear medicine. Nuclear technology has given doctors the use of instruments like CAT scans and MRIs which allows them to view internal body structures and therefore diagnose illnesses without invasive and harmful surgeries.

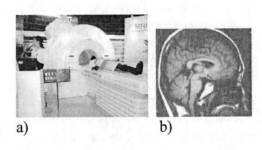

a)   b)

Figure 4-3
a) MRI instrument and b) Image generated from it

As the benefits of nuclear technology increase a better understanding of the nucleus and how it changes is apparent.

In the previous chapter we discussed the composition of the atomic nucleus in terms of two fundamental particles, the proton and neutron. The proton has an associated +1 charge and the neutron has a charge of zero; both have approximately the same relative mass of 1 amu. The total charge of the nucleus is given by the atomic number (number of protons), and the mass number is defined as the number of protons plus neutrons. We also know that atoms of a given element with different masses are called isotopes. During ordinary chemical reactions the properties of these isotopes are essentially identical; however, they behave quite differently when changes occur within the nucleus. For example, the two isotopes of carbon (C), $_{6}^{12}C$ and $_{6}^{14}C$, have very similar chemical properties (their chemical reactivities are identical), but, their nuclear properties differ considerably. The carbon-12 nucleus is very stable, whereas the carbon-14 nucleus decomposes spontaneously by means of radioactive decay. **Radioactivity** then, is the result of a natural change of an isotope of one element into an isotope of a different element. The discovery of radioactive isotopes led to technologies that were both beneficial and disastrous to human kind. So, no matter how you look at it, you have to agree that the tiny nucleus of an atom and the power it contains has had a tremendous effect on the world today.

In this chapter we will focus our attention on the nucleus and the changes it can undergo, and we will address some key questions such as:

When was radioactivity discovered? What are the products of radioactivity?
What is a nuclear change?
Why do some atoms undergo radioactive decay while others don't?
Why are some radioactive isotopes more harmful than others?
What are some useful applications of this process?

Discovery

The first type of nuclear reaction studied was that of natural radioactivity, in which the nucleus of an unstable isotope spontaneously decomposes. This phenomenon was discovered by a French scientist, Henri Becquerel, (Figure 4-1) as a result of an experiment he conducted with uranium salts.

Becquerel thought that exposing these salts to intense sunlight would produce high energy radiation similar to that of X-rays, (by this time were already discovered), which would in turn blacken photographic plates. His experiment proved successful. However, it would be proven later that the sunlight itself had zero influence on the production of the radiation.

Figure 4-4
Henri Becquerel

Becquerel concluded that the uranium salts spontaneously emitted this high-energy radiation whose existence had not previously been observed or even suspected. Furthermore, he found that the rate at which the radiation was emitted was proportional to the amount of uranium salts present. In other words, as the amount of uranium salts diminished, the amount of radiation emitted also decreased. Although Becquerel is credited for the discovery of natural radioactivity, it was a colleague of his, Marie Curie who isolated and studied several materials with radioactive properties for which she was awarded the Nobel Prize. The material Curie discovered showed much more intense radioactivity than uranium, and it was later named polonium, after Marie Curie's native country.

Figure 4-5
Marie Curie

Since the discovery of radioactivity, several groups of scientists set out to determine the identity of the radiation emitted. They eventually discovered that radiation from naturally occurring elements can be separated into three distinct parts: alpha rays or particles, beta rays or particles, and gamma rays.

Alpha rays consist of a stream of positively charged particles (**alpha particles**) that have a +2 charge and a mass of 4 amu. These particles are identical with the nucleus of the helium-4 isotope ($^{4}_{2}He^{+2}$). **Beta rays** are composed of a stream of negatively charged particles (**beta particles**) that have properties identical to an electron and are given the symbol, ($^{0}_{-1}e$). And finally **gamma rays (γ)** consists of electro-magnetic radiation of very short wavelengths (high frequency and high energy), making them the most damaging of all the radioactive particles.

Penetrating Ability of Radioactive Particles

Earnest Rutherford found that alpha rays could be easily stopped by thin pieces of paper, whereas beta rays could only be stopped by at least 0.5 cm of lead. In 1900 Paul Villard discovered the high-energy and extremely penetrating gamma ray, which showed characteristics of light waves and proved to be very damaging to human tissue. Gamma rays can only be stopped by a minimum of 10.0 cm of lead. See Figure 4-3

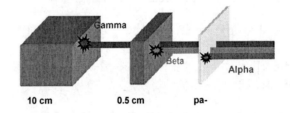

Figure 4-6
Penetrating power of the radioactive particles

Nuclear Reactions and Changes

As stated earlier, radioactivity is the result of a natural change from an isotope of one element into an isotope of a different element, which results in a **nuclear reaction**. The key players in such a reaction are the protons and the neutrons, which are collectively called **nucleons**. Therefore, during a nuclear reaction the number of nucleons remains the same but the identity of the element changes by emitting a radioactive particle or a ray.[4] Therefore, radioactive decay resulting in the release of an alpha

---

[4] Remember, the identity of an atom is based solely on the atomic number (the number of protons in the nucleus).; changing the number of protons therefore changes the atoms identity.

particle ($^4_2He^{+2}$) changes the identity of the atom. Atoms that undergo this type of decay are called **alpha emitters**. For example an isotope of the element radium (radium-226), decays by emitting an alpha particle and changes into a new element, the radon-222 isotope (equation 4-1).

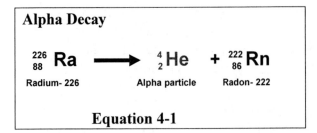

Alpha Decay

$$^{226}_{88}Ra \longrightarrow \, ^4_2He + \, ^{222}_{86}Rn$$

Radium-226   Alpha particle   Radon-222

Equation 4-1

Notice that the mass number on the left (226) equals the sum of the mass numbers on the right (4 + 222), as do the atomic numbers (88 = 2 + 86).

Example 4-2

Determine the product when Radon-222 undergoes alpha emission.
Answer:

Finding radon (Rn) on the periodic table gives its atomic number as (86). Therefore radon-222 has the following isotopic symbol

$$^{222}_{86}Rn$$

An alpha particle is given by:

$$^4_2He$$

Therefore:

$$^{222}_{86}Rn \rightarrow \, ^4_2He + \, ^A_ZX$$

The new element X, can be determined from A + 4 = 222 (the sum of the mass numbers on the right must equal the mass number (222) on the left). Therefore the mass number of the new element is (A=222 − 4 = 218). Likewise, the atomic number of the new element is found from: Z + 2 = 86 (the sum of the atomic numbers on the right must equal the atomic number (86) on the left). Therefore, the atomic number is (Z = 86 − 2 = 84). By referring to the periodic table we can identity the new atom as polonium (Po):

$$^{222}_{86}Rn \rightarrow \, ^4_2He + \, ^{218}_{84}Po$$

Notice that the alpha particles shown in Equation 4-1 and in the example above do not carry the +2 charge as described in the introduction. This is explained by noting the movement of electrons during the nuclear change. In the nuclear reaction shown in Equation 4-1, the atomic number of radium-226 changed from 88 to 86, because it loses two protons. This results in the atom having a net −2 charge because it now has 2 more electrons than protons. However, these two electrons are also removed causing the newly formed atom to be neutral. The two electrons are picked up by the alpha particle, which originally had a +2 charge and converted into a neutral species as well.

The nuclei of some atoms undergo emission of beta particles. These atoms are called **beta emitters**. For example, the radioactive decay of uranium-235 emits a beta particle ($^0_{-1}e$), and like alpha emitters changes atoms identity. This raises the question of how a nucleus can release a beta particle, which is essentially an electron. It has been determined that a neutron (^1_0n) is formed outside the nucleus from the combination of a proton (^1_1p) and an electron

48

($_{-1}^{0}e$). Therefore, the reverse reaction may occur inside the nucleus during beta decay.

$$_{0}^{1}n \rightarrow {}_{-1}^{0}e + {}_{1}^{1}p \quad \text{Equation 4-2}$$
neutron   electron   proton

When an atom undergoes beta decay a neutron in the nucleus changes into a proton, this increases the atomic number of the atom by one (Equation 4-3.)

Beta Decay

$$_{92}^{235}U \rightarrow {}_{-1}^{0}e + {}_{93}^{235}Np$$
uranium-235   beta particle   Neptunium-235

Equation 4-3

Example 4-3

Determine the product formed when lead-210 emits a beta particle.

Answer:

Finding lead (Pb) on the periodic table gives us its atomic number of (82). Therefore lead-210 has the following isotopic symbol

$$_{82}^{210}Pb$$

A beta particle is given by:

$$_{-1}^{0}e$$

Therefore:

$$_{82}^{210}Pb \rightarrow {}_{-1}^{0}e + {}_{Z}^{A}X$$

Again, the new element X, can be determined from the fact that the sum of the mass numbers on the right must equal the mass number (210) on the left. The mass number of the new element is (A = 210 − 0 = 210), which did not change. Likewise, the atomic number of the new element is found from: Z − 1 = 82, therefore Z = 82 + 1 = 83. Referring to the periodic table gives the identity of the new atom as Bismuth (Bi):

$$_{82}^{210}Pb \rightarrow {}_{-1}^{0}e + {}_{83}^{210}Bi$$

It is important to notice that the mass number of the new element formed during beta decay is the same as that in the original atom. This is because beta decay results in a neutron being converted to a proton, which does not change the mass number of the nucleus (number of neutrons plus protons).

After a nucleus emits an alpha or a beta particle it can still contain excess energy. This energy can be released in the form of highly energetic photons called **gamma rays (γ)**. These gamma rays can be powerful enough to penetrate a thick sample of lead or even a concrete wall. This explains much of the health damages associated with radioactive substances caused by gamma radiation, see Figure 4-4.

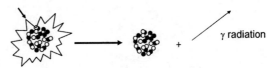

Figure 4-7
Gamma rays released from a nucleus having excess energy.

A fourth type of radioactive decay is positron emission. A **positron (β⁺)** is a particle equal in mass, but opposite in charge, to the electron. It is represented as ($_{+1}^{0}$e). Positron emission results in a decrease of atomic number by one without a change in the mass number. This is just the opposite of beta decay where the atomic number increases by one. For example, the isotope fluorine-18 decays by positron emission as shown in Equation 4-4.

Positron decay

$$^{18}_{9}\text{F} \quad \rightarrow \quad ^{0}_{+1}\text{e} \quad + \quad ^{18}_{8}\text{O}$$

fluorine-18   positron   oxygen-18

Equation 4-4

Example 4-4

Determine the product when phosphourus-31 emits a positron.
Answer:

Finding phosphorous (P) on the periodic table gives its atomic number as (15). Therefore, phosphourus-31 has the following isotopic symbol

$$^{31}_{15}\text{P}$$

A positron is given by:

$$^{0}_{+1}\text{e}$$

Therefore:

$$^{31}_{15}\text{P} \quad \rightarrow \quad ^{0}_{+1}\text{e} \quad + \quad ^{A}_{Z}\text{X}$$

Again, the new element X, can be determined from the fact that the sum of the mass numbers on the right must equal the mass number (31) on the left. Therefore, the mass number of the new element is (A = 31 – 0 =

31), shows no change. Likewise, the atomic number of the new element is found from: Z + 1 = 15, so Z = 15 - 1 = 14. By referring to the periodic table, we can identity of the new atom as silicon (Si):

$$^{31}_{15}\text{P} \quad \rightarrow \quad ^{0}_{+1}\text{e} \quad + \quad ^{31}_{14}\text{Si}$$

Stability of Atomic Nuclei

As mentioned earlier, some nuclei undergo spontaneous radioactive decay, but the majority of the naturally occurring elements are mixtures of stable isotopes of the elements. A **stable isotope** is one that does not spontaneously decompose into a different element. The stability of an isotope is related to several factors relating the ratio of neutrons to protons:

1. A nucleus is most stable when the number of neutrons is equal to or greater than the number of protons, with the exception of hydrogen-1 and helium-3.

2. For elements with low atomic numbers, the number of neutrons is very nearly equal to the number of protons in stable nuclei.

3. Nuclear stability is greater for isotopes that contain even numbers of protons and neutrons.

Using this information and plotting a graph of numbers of neutrons versus the numbers of protons, Figure 4-8, an apparent band of stability can be seen. This band (the darker dots) is the plot of the known stable nuclei. Nuclei outside this band will tend to decay in such a way that they change their neutron/proton ratio while moving closer to the stability band.

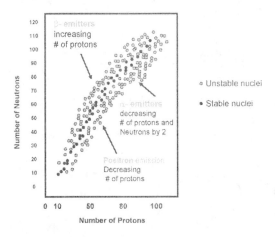

Figure 4-8
The darker dots represent the combinations of neutrons and protons in the known stable isotopes.

Beta decay usually occurs when a greater neutron to proton ratio exists. However, when a greater proton to neutron ratio exists, positron emission occurs. For unstable nuclei of elements with atomic numbers greater than 83, alpha emissions occurs decreasing the number of protons and neutrons by two and, as a result, they are moved closer to the stability band. Table 4-1 gives a summary of the different methods of radioactive decay.

Type of decay	Symbol	Charge	Mass	Change in atomic number	Change in mass number
Beta	$_{-1}^{0}e$	-1	0	+1	None
Positron	$_{+1}^{0}e$	+1	0	-1	None
Alpha	$_{2}^{4}He$	+2	4	-2	-4
Gamma	$_{0}^{0}\gamma$	0	0	None	none

Table 4-1
Summary of radioactive decay methods

Half Life

So far we have discussed radioactivity on a submicroscopic level, as it applies to a single atom. The study of radioactivity is very important in understanding the process of decay; however, it is not very practical when dealing with a large number of atoms. For example, if we could see the nucleus of single atom, based on its nuclear composition, we could determine if it would decay. The relative numbers of protons and neutrons determine an atoms stability and method by which it will decay; however, we could not tell when it would decay by simple observation. When an atom will decay is a completely random process and is generally uninfluenced by outside factors. With large numbers of atoms, radioactivity becomes much easier to predict. For example, the half life of a sample of radioactive atoms can easily be calculated. **Half life** is the period of time required for exactly one half of the number of atoms in the original sample to undergo a radioactive decay, forming a new element. Table 4-2 gives half lives for some common radioactive isotopes.

Element	Half-life
Uranium-238	4.5×10^9 years
Hydrogen-3 (tritium)	12.3 years
Carbon-14	5730 years
Indium-131	8.05 days
Copper-64	12.9 hours
Zinc-69	55 minutes

Table 4-2
Half lives for some common radioactive isotopes

Let's say a sample of matter contains exactly 10 billion atoms of the isotope potassium-40. The half life for this isotope is 1.3 billion years, which means that after 1.3 billion years 5 billion atoms of potassium-40 would have decayed, leaving 5 billion in the sample. From then it would take another 1.3 billion years for another 2.5 billion atoms to decay, leaving 2.5 billion atoms in the sample. A graphical representation of half lives is shown in Figure 4-9.

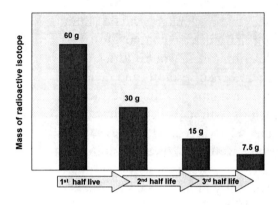

Figure 4-9
Graphical representation of half lives

We can calculate the fraction of the original isotope that remains after a given number of half lives from the following relationship:

$$\text{Fraction remaining} = \frac{1}{2^n}$$

Where n is the number of half lives.

Example 4-5

You have a fresh sample of 100 grams of zinc-69 isotope which has a half life of 55 minutes. How much zinc remains after 165 minutes?

Answer:

165 minutes / 55 minutes

= 3 half lives = n

Therefore:

$$\text{Fraction remaining} = \frac{1}{2^3} = \frac{1}{8}$$

$\frac{1}{8}$ x (100 grams) = 12.5 grams of zinc remains

Example 4-6

For a particular isotope with a half life of 120 years, how long will it take for an 80 gram sample to decay to 20 grams?

Answer:

80/2 = 40 grams = 1st half life

40/2 = 20 grams = 2nd half life

So, two half lives, each half life requires 120 years, therefore it will take

2 x 120 = 240 years

Applications of Radioactivity

Radio Isotopic Dating

The half life of some isotopes can be used to estimate the age of rocks and certain archeological artifacts. For instance, uranium-238 decays with a half life of 4.5 billion years. The initial products of this decay are also radioactive which eventually further decay to form the stable isotope lead-206. Scientist can measure the relative amounts of uranium-238 and lead-206 in a sample to determine its age. However, because its half-life is so large the use of uranium-

238 as a method for dating is only useful for materials older than 10 million years.

Carbon -14 Dating

Radiocarbon, or **Carbon-14, dating,** is one of the most widely used and most reliable dating methods for artifacts derived from plants and animals. It was developed by J. R. Arnold and W. F. Libby in 1949 and has revolutionized archaeology by providing a means for the establishment of world-wide chronologies. Since carbon-14 is a radioactive element with a half life of 5730 years and the most recent calculated age of the earth is 6.0 billion years old, how can there be enough carbon-14 on earth to be used for dating purposes? Carbon-14 is constantly being formed in the upper atmosphere when cosmic rays from the sun strike the stable nitrogen-14 atoms and converts them into the radioactive carbon-14. These newly formed carbon-14 atoms then combine with oxygen to form radioactive carbon dioxide. After being absorbed and used by plants the radioactive carbon dioxide gets into the food chain and the carbon cycle. Therefore, all living organisms, including plants, contain a constant ratio of carbon-14 to carbon 12 (the most stable form of carbon). Once the organism dies the carbon-14 exchange cycle stops, and any carbon-14 in the tissues of the organism decays forming nitrogen-14. Therefore, the change in the ratio of carbon-14 to carbon-12 is used for dating the artifact. Since the half-life of carbon-14 is only 5730 years this method can only be used on materials less than 70,000 years old.

Nuclear Medicine

Therapeutic and diagnostic uses of radioisotopes allow for very specific procedures involved in nuclear medicine. In therapeutic treatment, when rapidly growing cancer cells are subjected to doses of radiation, while minimizing the exposure to healthy tissue, their growth rates are drastically reduced and many times completely stopped. The treatment is primarily directed to the cancerous cells, but often times cause the patient to become nauseous and weak. Radiation therapy also reduces the levels of white blood cells and increases the likely hood of infections.

Radioisotopes are also used for diagnostic purposes to provide information about the type or extent of the illness. Radioactive technetium-99 is used in imaging of the brain, liver, bone marrow, kidney, lung and heart. Iodine-131 is used to determine the size, shape, and activity of the thyroid gland, as well as to treat cancer and hyperactivity of this gland. Small doses are used for diagnostic purposes while larger doses are used for treatment of thyroid cancer. Table 4-3 lists some radioisotopes and their medical uses.

Isotope	Name	Half-life hours	Uses
^{99}Tc	Technetium-99	6	Thyroid, brain, kidney
^{201}Tl	Thallium-201	21.5	heart
^{123}I	Iodine-123	13.2	thyroid
^{67}Ga	Gallium-67	78.3	Various tumors

Table 4-3
Some radioisotopes and their medical uses

Energy and Nuclear Reactions

We have seen that the radioactivities of unstable isotopes are very useful as archeological and medicinal tools. A more dramatic feature of the power contained in these isotopes is their release of immense energy by either nuclear fission or fusion. **Fission** is the splitting of heavy nuclei into smaller nuclei releasing large amounts of energy in the process. **Fusion** is the energy-releasing combination of light nuclei forming heavier ones. The power contained in the nucleus was determined by Albert Einstein, and at the age of 26 he had developed his special theory of relativity and worked out his famous equation:

$$E = mc^2$$

Where E = energy, m = mass and c = the speed of light. This equation suggests that mass and energy are interchangeable, or more specifically that mass is energy. According to Einstein, any reaction that liberates heat must lose mass. The amount of energy given off during the course of most reactions would correspond to a change in mass so small that it would not be able to be detected. For example, the combustion of 1000 grams of propane releasing 50,000 joules of heat would have a decrease in mass of only 5.0×10^{-13} kg (1 trillionth of a kg). Interestingly, the energy contained in a single potato chip weighing only ten grams would produce 9.0×10^{14} joules of heat (if all of its mass were converted to pure energy). This is enough energy to vaporize a block of ice the size of Manhattan. Of course, at the present we have no way of converting the mass of a potato chip to pure energy. However, it is easy to see that when large nuclei are split into smaller ones (fis-

sion), the energy released is tremendous and allows for a measurable change in mass. It has been calculated and measured that during nuclear fission the sum of mass of the two resulting fragments weighs less than the mass of the original fragment. This decrease in mass is of course responsible for the tremendous amount of energy released. One example of a fission reaction involves the splitting of a uranium-235 atom by a slow-moving neutron entering its nucleus. The addition of the neutron to the already unstable nucleus causes it break apart and release two new nuclei, two neutrons, and a large amount of energy. These released neutrons can potentially enter two other uranium nuclei causing them to split and release four neutrons and even more energy. If this process continues, a chain reaction occurs as shown in Figure 4-10. **Critical mass** is a term used to describe the size of the sample that is large enough to self-sustain a chain reaction. An explosion will occur if a critical mass of fissionable material is suddenly brought together, because the energy released from each fission reaction can not dissipate fast enough.

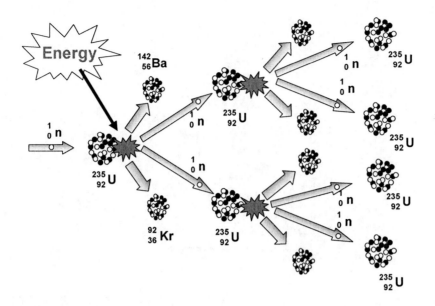

Figure 4-10
Illustration of a self sustaining nuclear chain reaction

Production of Electricity from Nuclear Reactions

In most electric power plants, water is heated and converted into steam, which drives turbine-generators and produces electricity. Fossil-fueled power plants produce heat by burning coal, oil, or natural gas. However, in a nuclear power plant, the fission of uranium atoms, as described in the previous section, provides the heat to produce steam for generating electricity. Several commercial reactor designs are currently in use in the United States. The most widely used design consists of a heavy steel pressure vessel surrounding a reactor core that contains the uranium fuel. The fuel is formed into cylindrical ceramic pellets about one-half inch in diameter, which are sealed in long metal tubes called fuel tubes. These tubes are arranged in groups to make a fuel assembly. A group of fuel assemblies forms the core of the reactor.

Heat is produced in a nuclear reactor when neutrons strike the uranium atoms and cause them to fission in a self-sustaining chain reaction. Control rods, which are made of materials that absorb neutrons, are placed among the fuel assemblies. When the control rods are pulled out of the core, more neutrons are available and the chain reaction speeds up, which produces more heat. When they are inserted into the core, more neutrons are absorbed, and the chain reaction slows or stops, and the amount of heat is reduced. Most commercial nuclear reactors simply use water to remove the heat created by the fission process. These are called *light water reactors*.

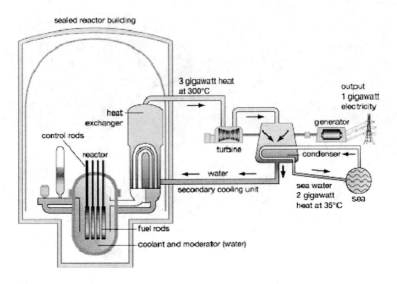

Figure 4-10
Diagram of a light water reactor

The water also serves to slow down, or "moderate" the neutrons. In this type of reactor (Figure 4-10), the chain reaction will not occur without the water to serve as a moderator.

Radioactive Waste

Every 18 to 24 months, nuclear power plants must shut down to remove and replace the spent uranium fuel. This **spent fuel** has released most of its energy through the fission process and has now become radioactive waste material.

All of the nuclear power plants in the United States produce about 2,000 metric tons of radioactive waste per year. Currently, this waste is stored at the nuclear plant at which it is generated, either in steel-lined, concrete vaults filled with water or in above-ground steel or steel-reinforced concrete containers with steel inner canisters, as shown in Figure 4-11. The Department of Energy is currently applying for a license to construct a permanent central repository at Yucca Mountain. If the license is granted, the repository could begin accepting waste by the year 2012.

Much of the equipment in the nuclear power plants becomes contaminated with radiation and will also become waste after the plant is closed. This will have to be stored and treated like the fuel waste for it too will remain radioactive for many thousands of years.

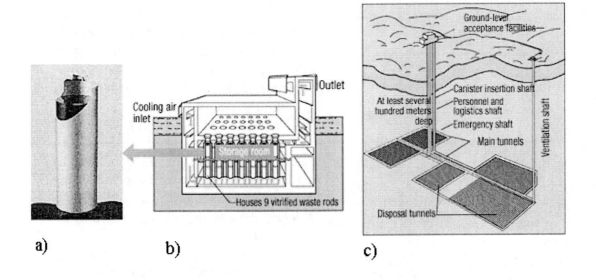

Figure 4-11
Example of radioactive storage facility, a) canisters for spent fuel rods b) high level radioactive storage facility at the surface c) underground radioactive waste facility.

Fusion and Energy

It is clear that fission reactors are a great source of energy; prior to their construction we depended primarily on the combustion of fossil fuels such as oil and coal to generate electricity. As most of us know, there exists a limit in the amount of these natural fuels. It has been estimated that we have about 50 years worth of oil, and 75 years worth of coal at the current rate of usage. As more fission reactors are constructed the amount of available fissionable material will also decrease. The fossil fuels have a strong environmental impact; they create problems like acid rain and an increase of CO_2 in the atmosphere, as we shall see in Chapter 7. Also, as mentioned in the previous section, nuclear power plants generate problems associate with the storage of radioactive waste. Furthermore, renewable energy sources, such as hydroelectric dams, solar power, and wind, are showing promise of handling an ever larger share of the energy demand but are resources that are spread out geographically and not capable of generating enough energy for the larger and more populated areas. Because of downfalls of these current sources of energy, the development of novel energy options is necessary. In doing so, we must pay special attention to security, environmental impact, and economical aspects. Thermonuclear fusion is one such option. Fusion reactions produce about the same amount of energy as fission on a per-gram basis but with fewer radioactive by-products. The by-products generated are much less hazardous than fission by-products and have much shorter half lives. As stated earlier fusion is the process by which two light nuclei are combined to form a larger one that releases a large amount of energy. When the two isotopes of hydrogen, deuterium ($_1^2H$) and tritium ($_1^3H$) are fused together at extremely high temperatures, helium-4 nuclei is formed and, 4.1×10^8 kcal of energy is released.

$$_1^2H + {}_1^3H \rightarrow {}_2^4He + {}_0^1n + 4.1 \times 10^8 \, kcal$$

Equation 4-5

Notice from Equation 4-5 that an extreme amount of energy is released with a minimum amount of waste. So the question is why we aren't using fusion instead of fission. Could it be that there is not enough fusible material on the planet to make it worth while? Actually, deuterium is an excellent fuel for fusion, and is very abundant in the oceans. In every liter of ocean water there is 1.0×10^{22} atoms of deuterium, which if used in a fusion reaction, would release enough energy to provide electricity to every home and business in the United States for 50 years. Obviously, fuel is not the problem. The problem is getting fusion reactions started. Unlike fission reactions, fusion requires a large amount of energy, usually in the form of heat (in the 100 million degree range) and pressure, in order to be initiated. Since most materials are not capable of handling such extreme temperatures, confining the fusible material is a problem. Using a plasma form of the fuel may solve this problem. **Plasma** forms when atoms are stripped of their electrons, which results in positively charged nuclei. This plasma can be held in place or contained in a magnetic field where fusion could be allowed to take place. At the moment, scientist have not been able to generate the heat needed to initiate and sustain a fusion reaction, however, when they do, it will provide for a very abundant and low-cost energy source for all of us.

Radiation and Food

Food irradiation is a promising new technology used in the elimination of disease-causing germs from foods. Like the pasteurization of milk and pressure cooking of canned foods, irradiation of foods can kill bacteria and parasites that would otherwise cause disease. The effects of irradiation on the food and on animals and people eating it have been studied extensively. These studies show clearly that when irradiation is used as approved:

- disease-causing germs are reduced or eliminated from the food
- the food does not become radioactive
- dangerous substances do not appear in the foods
- the nutritional value of the food is essentially unchanged

Therefore, irradiation has been shown to be a safe and effective technology that can prevent many food related diseases.

Chapter 4 Exercises

1. Who first discovered radioactivity?

   _____

2. Which type of radiation is the most penetrating?

   _____

3. Which is not generally associated with nuclear reactions?

   a. outer electrons   b. atomic number   c. atomic mass   d. the nucleus

4. Which type of radiation has a charge of +1 and a mass of zero?

   a. positron   b. beta   c. alpha   d. neutron

5. What type of particle is emitted when a U-235 decays to Np-235?

   a. alpha particle   b. beta particle   c. neutron   d. helium nuclei

6. Stable nuclei (that is, non-radioactive nuclei) have mass numbers that are

   a. equal to the atomic number   b. twice as large as the atomic number or even larger
   c. ten times the atomic number   d. less than twice as large as the atomic number

7. What is the iodine-123 isotope used for

   _____

8. Which is the opposite of an electron?

   a. α-particle   b. positron   c. neutron   d. β-particle

9. Define nucleons?

_____

10. If the half-life of an element is 5 days and 100 grams of that element is initially available, how many grams of the element are present after 15 days?

_____

11. In the symbol ^{238}U, the 238 is

_____

12. The product of the beta ($_{-1}^{0}e$) decay of thorium-234 (^{234}Th) is

_____

13. The product of the alpha ($_{2}^{4}He$) decay of uranium-238 is

_____

14. What is produced when Be-9, decays by positron emission?

_____

15. How many atoms of zinc-69 remain after a sample of 500,000 zinc-69 atoms decays for a period of 180 minutes? The half-life time for zinc-69 is 90 minutes.

_____

16. Which of the following is an effect of food irradiation?

_____

_____

_____

_____

17. When the nucleus of an atom of uranium-238 emits an alpha particle, the mass number will

a. decrease by 4 units   b. decrease by 2 units   c. increase by 4 units   d. increase by 2 units

18. Which types of emission occurs when a greater neutron to proton ratio exists?

_____

19. When a radioactive sample decays for 3 half-lives the amount remaining will be _____ of the original.

a. 1/4   b. 1/2   c. 1/8   d. unpredictable

20. In terms of nuclear stability, what atom is the most stable?

a. Fe   b. H   c. U   d. element 109

21. Define fission?

_____

_____

_____

_____

_____

_____

_____

22. Which component of a nuclear reactor slows the chain reaction?

a. shield   b. heat-transfer fluid   c. control rods   d. uranium fuel

23. Which is true about fusion?

   a. very high temperatures are required   b. occurs in a plasma of charged particles
   c. produces

   fewer radioactive byproducts   d. all statements are true _____

24. The missing nucleus in the equation shown below is

   $_{90}^{230}\text{Th} \rightarrow {}_2^4\text{He} + ?$

   _____

25. Tritium, H-3, and hydrogen, H-1, are

   a. fission reactants   b. fission products   c. fusion reactants   d. fusion products

26. The splitting of a nucleus by a slow moving neutron is called

   _____

27. A major, unsolved problem with nuclear energy is

   _____
   _____
   _____
   _____
   _____

28. The fusion of nuclei requires

   a. a critical mass   b. plasma at extremely high temperatures   c. heavy nuclei   d. slow
   moving (thermal) neutrons

29. What produces the tremendous energy of a nuclear fission reaction?

   a. some mass being changed into energy   b. the splitting of the nucleus

   c. the extra neutrons produced   d. disintegration of the containment building

Chapter 5

Bonding of the Atoms

Atoms of almost every element have the ability to combine with other atoms to form more complex structures. The forces of attraction that bind them together are called chemical bonds. To truly understand chemistry, the nature and origin of the chemical bond is important, since the basis for all chemical reactions is the forming and the breaking of bonds and the resulting energy changes. The two main types of bonding are covalent and ionic, which both result from the interaction of the electrons between atoms. Chemical bonds determine the three dimensional shape and physical properties of molecules. Whether a substance is a solid, liquid or a gas has a lot to do with its structure. In this chapter we will consider the types of bonding in compounds and molecules while addressing some key questions such as:

What are ionic bonds?

What are the names and formulas for
 ionic compounds?

What are polyatomic ions?

What are covalent bonds?

What are the names of covalent
 compounds?

How do we predict the shape of
 molecules?

What are polar and non-polar bonds?

What are the properties on ionic and covalent compounds?

What are intermolecular forces?

What are the states of matter?

What makes water unique?

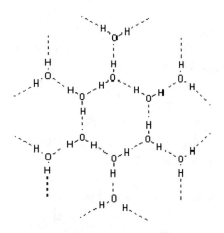

Figure 5-1

Covalent and hydrogen bonding in water

Ionic Bonding

As was mentioned in Chapter 3, atoms tend to combine in such a way to achieve a noble gas electronic configuration – the **Octet rule**. The name *octet* derived from the fact that the noble gases all have *eight* valence electrons. There are, however, many more reasons for atoms to combine, like the lowering of their energies, but the octet rule will suffice for our current discussion.

We saw in Chapter 3, that atoms have a tendency to either gain or lose electrons to form ions. Cations are formed when atoms lose electrons and anions are formed when atoms gain electrons. Their abilities to gain or lose electrons are related to their ionization energy (the energy needed to remove an electron from an atom). This tendency is also encouraged by the atoms desire to achieve a noble gas electronic configuration. When atoms of elements come into contact, one or more electrons can be transferred between them which form of a cation and an anion. The direction of transfer depends on which atom has the stronger desire to receive an electron. Lewis dot symbols can be used in determining the direction of electron transfer. In **Lewis dot symbols,** the valence electrons represented by dots are placed around the symbol until they are all used or until all four sides are occupied. Figure 5-2 shows the Lewis dot symbols for the representative elements.

IA	IIA	IIIA	IVA	VA	VIA	VIIA	VIIIA
·H							He:
·Li	·Be·	·B·	·C·	·N·	·O·	:F·	:Ne:
·Na	·Mg·	·Al·	·Si·	·P·	·S·	:Cl·	:Ar:
·K	·Ca·	·Ga·	·Ge·	·As·	·Se·	:Br·	:Kr:
·Rb	·Sr·	·In·	·Sn·	·Sb·	·Te·	:I·	:Xe:
·Cs	·Ba·	·Tl·	·Pb·	·Bi·	·Po·	:At·	:Rn:
·Fr	·Ra·						

Figure 5-2
Lewis dot symbols for the representative elements

Using the Lewis dot symbols shown in Figure 5-2 and the octet rule, it is easy to see which elements want to gain electrons and which ones want to lose them. For example, chlorines symbol is surrounded by seven dots, representing its seven valence electrons, therefore it only needs one more electron to satisfy the octet rule. On the other hand, sodium has only one dot representing its valence electron, and has to either gain seven or lose one electron to achieve a noble gas configuration. Of course its easier for sodium to lose, or give up, its one valence electron. Therefore if sodium and chlorine were to come into contact the sodium atom would transfer its one electron to chlorine and become a positively charged ion called a *sodium ion*. Chorine accepting the electron would become a negatively charged ion called a *chloride ion,* as shown in Equation 5-1. These two ions of opposite charges are now mutually attracted to one another by **electrostatic forces** (forces between particles caused by their differences in electric charges) also called the **ionic bond** and they form an **ionic compound**.

$$Na· + :\ddot{C}l· \longrightarrow Na^+ + :\ddot{C}l:^-$$
sodium atom   chlorine atom   sodium ion   chloride ion

$$\downarrow$$

NaCl
Sodium chloride compound

Equation 5-1

The attraction of the oppositely charged ions allows them to settle into a regular crystalline structure, or **crystal lattice,** and form ionic compounds as shown in Figure 5-3.

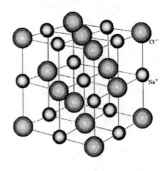

Figure 5-3
Crystal structure of NaCl

Neutral ionic compounds formed from metal ions (cations) and nonmetal ions (anions) are commonly called **salts**. Sodium chloride (NaCl), or table salt, is an example of a neutral ionic compound of a metal ion (Na^+) and a nonmetal ion (Cl^-), combining to form a salt. Salt compounds are "neutral" because the charges on the individual ions cancel each other and cause a net charge of zero (the sodium ion has a charge of +1 and chloride ion -1, and so +1 and -1 equals zero).

Predicting Formulas of Ionic Compounds or Salts

It is possible to figure out or predict the formulas of ionic compounds, (salts) by considering the charges of the ions formed from the elements involved. The number of electrons gained or lost by atoms upon reaching a noble gas configuration determines the charges on their respective ions. The common ionic charges for the representative elements are given in Table 5-1.

IA	IIA		IIIA	IV	VA	VIA	VIIA
H^+							
Li^+				$C^{\pm4}$	N^{3-}	O^{-2}	F^-
Na^+	Mg^{2+}		Al^{3+}	$Si^{\pm4}$	P^{3-}	S^{-2}	Cl^-
K^+	Ca^{2+}					Se^{-2}	Br^-
Rb^+	Sr^{2+}					Te^{-2}	I^-
Cs^+	Ba^{2+}						

Table 5-1
          Common ions formed for the representative elements

Notice the elements of groups IA, IIA and IIIA form cations while groups VA, VIA and VIIA form anions. From the table it is clear that the elements in groups IA – IIIA lose one, two and three electron(s) respectively while those in groups VA – VIIA gain three, two and one electron(s) respectively. The reason why the groups IVA's ionic charges, given in Table 5-1, are shown as (±4) is because, depending on which elements they combine with, these elements can either gain four or lose four electrons.

It is now an easy task to predict the neutral formula for any combination of cations and anions. For instance when the cation of magnesium (Mg^{2+}) combines with the anion of chlorine (Cl^-), the neutral formula must be $MgCl_2$. Two chloride ions having a charge of -1 are each required to cancel the +2 charge of the magnesium ion.

An easy way for determining the numbers of each ion needed in the compound formula is to use the *charge cross over method*. This method uses the absolute value of each ion and places it as a subscript on the opposing element. For example calcium forms the calcium +2 ion (Ca^{+2}) where bromine forms the bromine -1 ion (Br^-) (notice that when the charge is ±1 the number one is omitted). Using the cross over method the ions are first placed side by side (the metal is always written on the left and the nonmetal on the right) with their charges shown. Then underneath write the ions without their charges; the absolute value of the cations charge (+2) is written as a subscript on the nonmetal (Br) and the absolute value of the anions charge (-1) is written as a subscript on the metal (Ca) as shown below:

65

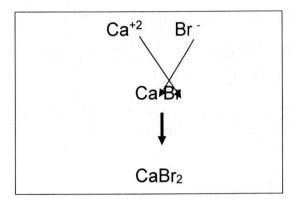

Naming Binary Ionic Compounds

Both names and formulas are used to identify compounds. During the early days of chemistry many different methods were used for the naming of compounds. As the numbers of newly isolated and developed compounds increased an organized system for naming was needed. **Chemical nomenclature** is the system that scientists use for naming ionic and molecular compounds. When naming ionic compounds consisting of two atoms (**binary compounds**), where one is a metal (cation) and the other a nonmetal (anion), use the elemental name of the cation followed by a modified name of the anion. Change the ending of nonmetal or anion name to include **–ide**; for instance chlorine formed as an anion would be named as the chloride ion. So, a compound formed from the combination of sodium and chlorine would is named sodium chloride.

Example 5-1

Determine the compound formulas when: a) sodium combines with oxygen and b) calcium combines with nitrogen

Answer

a) Sodium (Na) has an ionic charge of +1 and oxygen (O) is -2.

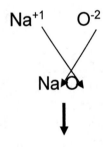

Compound Formula

b) Following the same method the compound formula for calcium and nitrogen is:

$$Ca_3N_2$$

Example 5-2

Determine the name and formula formed from the combination of aluminum and oxygen.

Answer

Keeping its elemental name, aluminum (the cation), is named first and the anion oxygen is named after the ending is changed to include –ide (oxide) making the name **aluminum oxide.** Using the crossover method and noting that aluminum has a +3 charge and oxygen a -2 charge gives the name and formula as:

Aluminum oxide Al_2O_3

Polyatomic Ions

Polyatomic ions are compounds formed from more than one element having an overall charge. For instance, the nitrate ion is formed from one nitrogen atom and three oxygen atoms, which has an overall charge of -1 and is written as (NO_3^{-1}). In ionic compounds polyatomic ions are treated as single entities and combine with other ions in the same manner as do monatomic ions. The formulas and names of compounds containing polyatomic ions are determined the same as with binary compounds; however, the name of the polyatomic ion is not changed. For instance, the compound formed from a potassium ion and a nitrate ion is potassium nitrate and the formula is (KNO_3). For clarity, a parenthesis is used if more than one polyatomic ion is needed in a compound formula. Aluminum nitrate, an ionic compound (salt) formed from the aluminum ion and a nitrate ion is named aluminum nitrate and has the compound formula;

$$Al(NO_3)_3$$

Notice, without using a parenthesis the formula would be very confusing and written as:

$$AlNO_{33}$$

Table 5-2 gives the names and formulas for some common polyatomic ions.

Example 5-3

Determine the name and formula of the salt (ionic compound) formed from the calcium ion and the polyatomic anion perchlorate.

Answer

The combination of Calcium (Ca^{2+}) and perchlorate (ClO_4^-) is simply named calcium perchlorate. Using their charges and the crossover method, the formula would be written as:

$$Ca(ClO_4)_2$$
Potassium perchlorate

Name	Formula	Name	Formula
Acetate	$CH_3CO_2^-$	Nitrate	NO_3^-
Carbonate	CO_3^{2-}	Nitrite	NO_2^-
Bicarbonate	HCO_3^-	Permanganate	MnO_4^-
Chlorate	ClO_3^-	Phosphate	PO_4^{3-}
Perchlorate	ClO_4^-	Hydrogen phosphate	HPO_4^{2-}
Chromate	CrO_4^{2-}	Dihydrogen phosphate	$H_2PO_4^-$
Cyanide	CN^-	Sulfate	SO_4^{2-}
Dichromate	$Cr_2O_7^{2-}$	Bisulfate	HSO_4^-
Hydroxide	OH^-	Sulfite	SO_3^{2-}

Table 5-2
Names and formulas of some polyatomic ions

Covalent Bonds

Nonmetallic atoms combine with each other to form compounds for much of the same reason that metals and nonmetals combine, in that they try and achieve a noble gas configuration. However, the manner in which they combine is quite different. Ionic bonds are formed when one atom transfers electron(s) to another atom, which produces oppositely charged particles that are attracted toward one another. In a **covalent bond**, atoms "share" some or all of their valence electrons in such a manner as to achieve a noble gas configuration about each atom. Using the Lewis Dot symbols from Figure 5-2, we can see that when two hydrogen atoms share their single valence electrons they form the hydrogen molecule having the molecular formula (H_2), which results in both atoms having two electrons about them, as does the noble gas helium. This form of representation is called the **Lewis Dot structure**.

•H + •H ⟶ H**:**H

hydrogen atoms   **Lewis dot structure** for a covalently bonded hydrogen molecule (H_2)

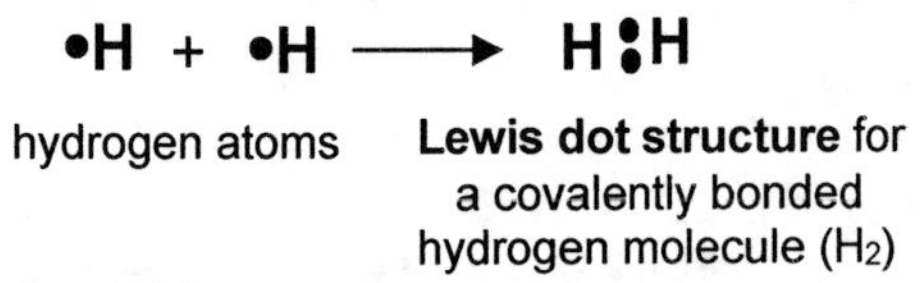

Furthermore, while compound formula is used in ionic compounds, the term **molecular formula** is used when referring to covalently bonded compounds.

Single, Double and Triple Covalent Bonds

When atoms combine to form a single **covalent bond**, a bond formed when two atoms share two electrons, the bonding electrons (or **pair of electrons**) can be represented by a single line drawn between the two atoms, as shown in Figure 5-4.

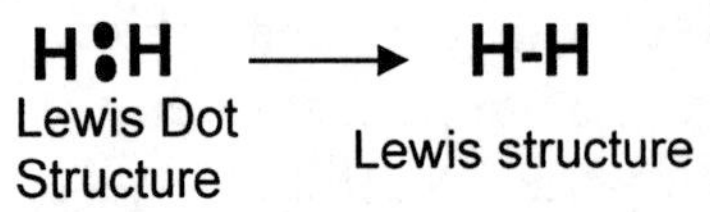

H**:**H ⟶ H-H
Lewis Dot   Lewis structure
Structure

Figure 5-4
    Lewis Dot structure and Lewis structure

When bonding electrons are represented by lines the structure is called a Lewis structure instead of the Lewis Dot structure.

Example 5-4

Draw the Lewis Dot structure and the Lewis structure for the molecular compound formed from the hydrogen atom and the fluorine atom.

Answer

From the Lewis Dot symbols we have;

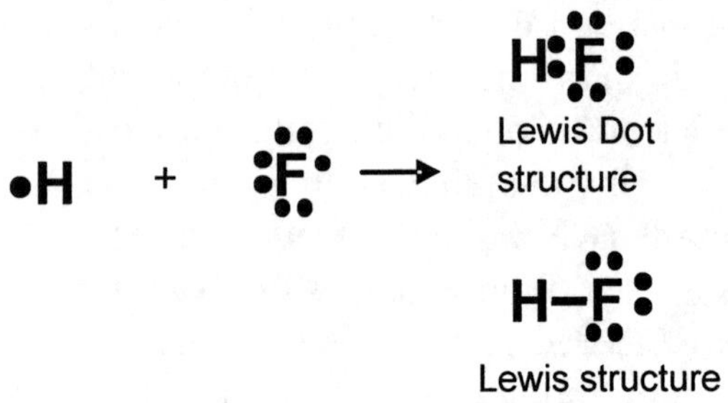

•H + **:**F**:** ⟶ H**:**F**:**
Lewis Dot structure

H-F**:**
Lewis structure

Before they combine as shown in the example above, hydrogen has one valence electron while fluorine has seven. After combining, however, fluorine has eight electrons surrounding it, and hydrogen has two. In both cases the atoms have achieved a noble gas configuration; hydrogen becomes like helium and fluorine like neon.

A **double bond** results when two atoms share four electrons. For instance, in order for oxygen, with only six valence electrons, to obtain a noble gas configuration when combining with another oxygen atom, four electrons must be shared between the two atoms; this gives both oxygen atoms eight electrons and result in the formation of a double bond as shown in Figure 5-5.

68

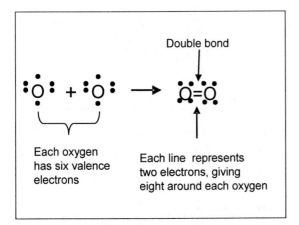

Figure 5-5
   Formation of a double bond when two oxygen atoms combine in such a way as to achieve a noble gas configuration

A **triple bond** forms when six electrons are shared between two atoms. This can be seen when two nitrogen atoms combine to form the nitrogen molecule (N₂), as shown in Figure 5-6.

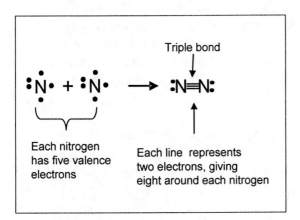

Figure 5-6
   Formation of a triple bond when two nitrogen atoms combine in such a way as to achieve a noble gas configuration

Covalent bonding also applies to molecules with more than two atoms such as a class of compounds called hydrocarbons (Chapter 9), molecules containing only carbon and hydrogen. For example, methane, a compound containing one carbon and 4 hydrogen atoms (CH₄), will covalently bond in the same manner as do the previous examples. Carbon has four valence electrons; therefore, four more are needed to satisfy the octet rule. Combining with four hydrogen atoms, each with one valence electron satisfies carbon's need for eight electrons, as shown in Figure 5-7.

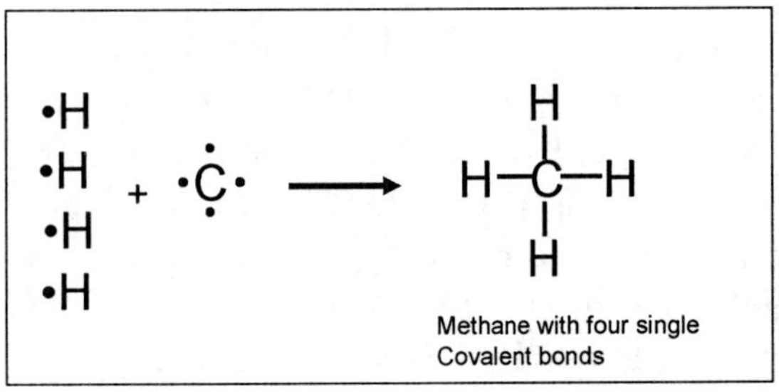

Figure 5-7
   One carbon atom and four hydrogen atoms will share electrons to achieve a noble gas configuration

When ethylene (C₂H₄) is formed from two carbon atoms and four hydrogen atoms, a double bond must form in order for each atom to have a noble gas configuration (Figure 5-8).

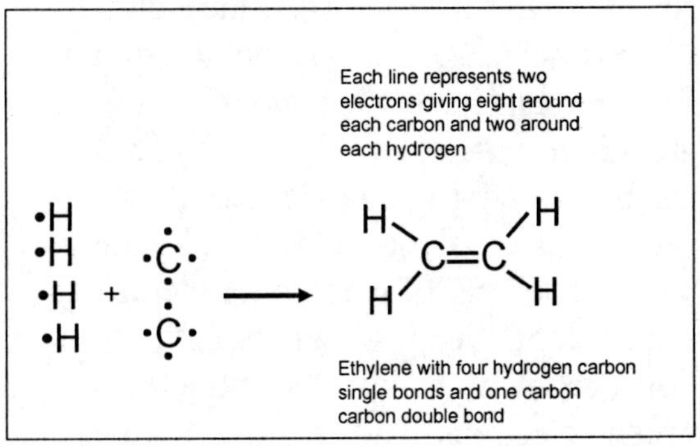

Figure 5-8
   Ethylene with four carbon-hydrogen single bonds and one carbon-carbon double bond

Bond Energies

Although covalent bonds can form between many different types of atoms not all covalent bonds are identical and as a result, have different bond energies. **Bond energy** is defined as the energy required breaking 6.02×10^{23} bonds between a specified pair of atoms. Single, double, and triple bonds clearly differ in the number of electrons used and, as a result, the energy required to break a triple bond is greater than that of a double or single bond. Think of the number of electrons in a bond as the number of nails holding two boards together; six nails (as in a triple bond) will hold stronger than four nails (double bond) which in turn holds stronger than two nails (single bond). To extend the analogy, the types of boards used will also affect the strength of the binding between them. For instance, balsa (a very fragile wood) would be easier to separate than cherry (very hard wood). Likewise, the types of atoms used in covalent bonding affects the strength of the bonds. For example a carbon-carbon double bond is stronger than a nitrogen-nitrogen double bond. Furthermore, as the number of bonding electrons increases, the distance between the bonding atoms decreases (i.e. A single bond is longer than a double bond which is longer than a triple bond). Table 5-3 gives bond energies and lengths for a few types of covalent bonds. Although there are a number of reasons for the difference in bond lengths and strengths of various covalent bonds, the attractions between electrons and nuclei of the bonding atoms has a large influence and is worth further discussion.

Bond type	C-C	C=C	N-N	N=N
Bond length (nm)	0.154	0.134	0.140	0.124
Bond energy Kcal/mol	83	146	40	100

Table 5-3
 Bond lengths and energies for a few different types of bonds

Electronegativity

Electronegativity is the ability of an atom during bond formation to attract electrons from its bonding partner. The higher the electronegativity, the stronger is the atom's electron attracting ability. You will recall that the nonmetals tend to gain electrons while metals tend to lose electrons during ion formation. Therefore, we can say that nonmetals have higher electronegativities than metals. It turns out that as you move across the periodic table from left to right, electronegativities increase, and as you move down a column they decrease. See Figure 5-9.

Electronegativity Trends

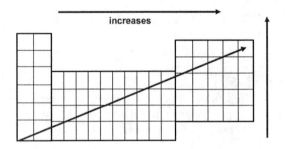

Figure 5-9
 Electronegativity trends for atoms on the periodic table

Based on these patterns, fluorine is the most electronegative element, which means it has the strongest tendency to attract electrons toward itself. Francium, on the other hand, has the lowest electronegativity, or the weakest tendency to attract electrons. We can now use the differences in electronegativities to discuss the attraction the atoms have to the electrons in a covalent bond. When a covalent bond is formed between two identical atoms, the electrons are shared equally between them because there is no tendency for either atom to attract the electrons toward itself more. However, when the atoms are different, a **polar covalent bond** is formed in which the electrons are not shared equally. This unequal sharing of electrons results in a buildup of a negative charge on the side of the bond closest to the more electronegative atom. This creates a charge distribution much like those found in magnets, hence the name polar, and gives rise to a North and South Pole. For example, compare the bonds formed in the hydrogen molecule (H_2) with those of the hydrogen fluoride (HF) molecule. In the hydrogen molecule the electrons in the bond are shared equally, so it is non-polar, in hydrogen fluoride, the electrons are attracted to the more electronegative fluorine, which produces a polar bond between the hydrogen and fluorine atoms, with a partial negative charge (δ^-) on the fluorine and partial positive charge (δ^+) on the hydrogen. Figure 5-10 shows the charge distribution for the bonds in hydrogen and hydrogen fluoride.

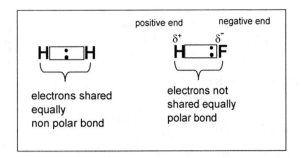

Figure 5-10
Charge distribution in the hydrogen molecule and the hydrogen fluoride molecule

The strength of a polar bond, or its **polarity**, is based on the differences in electronegativities of the atoms involved in bonding. In general, the greater the difference in electronegativity between the bonding atoms, the stronger the polar bond formed. For instance, the polarity of the bond between nitrogen and fluorine is not as strong as that formed between hydrogen and fluorine. The differences in electronegativities can be inferred from the atoms' location on the periodic table and the electronegativity trends, shown in Figure 5-9.

Example 5-5

Which bond is more polar?

N-O or B-O

Answer

Since boron and oxygen are farther apart on the periodic table we can predict that the B-O bond is more polar.

Nomenclature of Binary Covalent Compounds

Binary covalent compounds are covalently bonded compounds composed of two types of elements. As was mentioned earlier, these compounds are primarily formed from two different non-metal elements. Naming of these compounds is similar to that of the binary salts. When naming salts we always first name the metal then the nonmetal with an alternate ending of –ide. If we only consider electronegativities of the atoms involved then we can say that the less electronegative element is named first (as it appears on the periodic table) and followed by an altered name of the second element. If more than one atom of a specific element is found in the compound, a prefix representing the number of the atoms is used. For example, the compound formed from one carbon atom and two sulfur atoms is (CS_2) and is named carbon disulfide. Although there are three atoms in the molecule of carbon disulfide it is considered a binary compound since it is made of only two types of elements. Carbon is named first because it is found in a lower numbered group (group IVA) than sulfur (group VIA) and, therefore, has a lower electronegativity. Because sulfur has a higher electronegativity, it is named second, and its name is changed to *sulfide*. Since there are two sulfur atoms, the prefix di- is added to sulfide. Table 5-4 gives the prefixes used when more than one atom of a specific element is found in a binary covalent compound.

Number of atoms	prefix	Number of atoms	prefix
1	mono-	6	hexa-
2	di-	7	hepta-
3	tri-	8	octa-
4	tetra-	9	nona-
5	penta-	10	deca-

Table 5-4
Prefixes used representing the number of atoms found in a binary covalent compound

The prefix "mono" meaning "one" is used only when it refers to the second element in a compound, for instance, we do not name the molecule (CS_2) *monocarbon disulfide*. However, the compound formed from two nitrogen atoms and one oxygen atom is named dinitrogen monoxide (N_2O).

Example 5-6

Name the compounds formed from the following sets of elements.

a) two oxygen's and one sulfur
b) one carbon and one oxygen
c) one silicon and four fluorine's

Answers

a) oxygen is more electronegative and is therefore named second as oxide. Since there are two oxygen atoms, the prefix di- is used. Therefore the name is;

a) sulfur dioxide

The others were determined in the same manner.

b) carbon monoxide
c) silicon tetrafluoride

72

3-Dimensional Structures of Molecules

The 3-D shapes of molecules are predicted based on the premise that groups or pairs of electrons surrounding an atom are repelled as far apart as possible from each other, which minimizes the overall energy of the molecule. These electrons can be bonding or non-bonding pairs of valence electrons. This prediction method is often referred to as the **valence shell electron pair repulsion** theory (VSEPR). The basis of the VESPR theory is that groups of electrons arrange themselves about the central atom in such a way that lowers the molecules overall energy. In other words they will get as far apart as possible. The number and types of electron pairs or groups are determined from a Lewis structure. For simplicity we will only consider the arrangement of two, three and four groups surrounding a central atom. The **central atom** is the atom in a molecule to which all other atoms are bonded. From a molecular formula the central atom is usually the "odd" atom or the atom that is different from the rest. For instance the central atom in methane (CH_4) is carbon. Once a Lewis structure is correctly drawn the 3-dimensional shape of the molecule can be predicted. The steps involved in drawing a correct Lewis structure are:

1. Draw a skeleton structure where the atoms are arranged about the central atom in a symmetrical manner and connected by a single line.

2. Add or count up the total number of valence electrons available.

3. Subtract two electrons from the available valence electrons, from step 2,

for every line drawn in the skeleton structure.

4. Determine the number of electrons needed for each atom to satisfy the octet rule (having eight electrons surrounding them, remembering however, that hydrogen can only have a maximum of two electrons surrounding it).

5. If an exact number of electrons are available (to satisfy the octet rule) add enough electrons around the four areas of the atom until they each have eight. If there are not enough electrons available to satisfy the octet rule, add another line (making a double bond) for every two electrons needed to the skeleton structure and continue from step 3.

Let's take, for example, ammonia (NH_3);

Step 1:
Skeleton structure

$$H-\underset{\underset{\displaystyle H}{|}}{N}-H$$

Step 2: Nitrogen is in group VA, therefore, it has five valence electrons, and hydrogen is in group IA and has only one, but since there are three hydrogens we have three valence electrons. So, the total number of valence electrons available is eight.

Step 3: There are three lines drawn in the skeleton structure so we must subtract six electrons from the eight available, leaving two electrons.

Step 4: Since a single line in the skeleton structure represents two electrons, each hydrogen has two electrons surrounding it, which is the maximum number it can have. However, because nitrogen wants eight and only has six in the skeleton structure, it needs two more electrons. So the total number of electrons needed to satisfy all atoms in the molecule is two.

Step 5: Since the number of electrons needed is equal to the number available we can place the electrons around each atom satisfying the octet rule:

Correct Lewis structure

Example 5-7

Determine the Lewis structure for carbon dioxide (CO_2)

Answer:

Step 1:
Skeleton structure   O—C—O

Step 2:   C = 4 valence electrons
              O x(2) = 12 valance electrons
Total available = 16 valence electrons

Step 3: Subtract 4 for each line in the skeleton structure leaving 12 valence electrons available.

Step 4: Each of the atoms wants eight electrons around it. At the moment, carbon has four (two lines) and each oxygen has two (one line); carbon needs four electrons and each oxygen needs six. The total electrons needed are sixteen.

Step 5: We are four electrons short since only twelve electrons are available. We have to add two more lines to the skeleton structure making two double bonds:

O=C=O

Step 3 again: Now that we have added two more lines to the skeleton structure we have to subtract four more electrons from the available twelve, leaving only eight.

Step 4 again: Carbon now has eight electrons around it (4 lines) so its octet is satisfied; each oxygen has four (2 lines) so it needs four more. Therefore, the total electrons needed are eight.

Step 5 again: We need eight electrons and have eight available. Adding four electrons (or two pairs of electrons) to each oxygen atom gives the correct Lewis structure as:

Ö=C=Ö

We will consider two types of 3-dimensional shapes; **electron pair shape**, those including the un-bonded electron pair connected to the central atom, and **molecular shape**, those ignoring the un-bonded electron pair connected to the central atom. Let's assume that the central atom has four areas where groups of electrons can be placed: the top, bottom, left and right sides as shown in Figure 5-11. From a correctly drawn Lewis structure you can determine how many of these sides are filled with electron groups. For instance, from the example for ammonia (NH_3), we can see that all four areas contain electron groups.

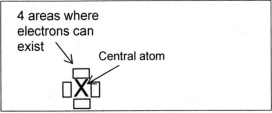

Figure 5-11
Four areas surrounding the central atom where electron groups can reside

In Example 5-7, the Lewis structure for CO_2 shows only two areas occupied by electron groups. According to VSEPR, electron groups around a central atom will arrange themselves as far apart as possible. Table 5-5 shows us the electron group separation and electron pair shapes when two, three, and four areas are occupied around a central atom.

Areas occupied	Maximum separation	Electron pair shape	Example
2	180°	Linear	CO_2
3	120°	Trigonal planar	H_2CO BCl_3
4	109.5°	Tetrahedral	CH_4 $SiCl_4$

Table 5-5
VSEPR results for a central atom with 2, 3, and 4 areas occupied by electron groups

Using Table 5-5 we can predict that ammonia (NH_3) will have an electron pair shape, or E.P. shape, of a tetrahedral, and that carbon dioxide's (CO_2) E.P shape will be linear. More examples of E.P shapes are given in Figure 5-12.

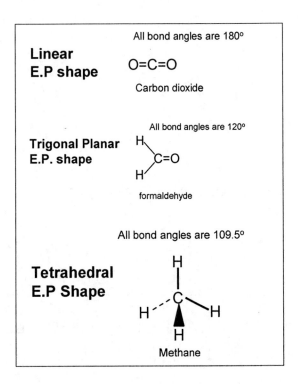

Figure 5-12
Examples of linear, trigonal planar, and tetrahedral E.P shapes

Example 5-8

What is the E.P shape for H_2O?

Answer:

Correct Lewis structure for H_2O H–Ö–H

There are 4 areas occupied around the oxygen atom. According to Table 5-5, the E.P shape is tetrahedral.

Molecular shapes are then derived from E.P shapes where the un-bonded electron pairs are ignored in the 3-dimensional structure. For example the E.P shape of ammonia is:

If we then ignore the un-bonded pair we have:

The bond angles are still 109.5° as in a tetrahedral, but because the electron pair is still there, we are not showing it. We find the molecular shape of ammonia to be trigonal pyramid. Table 5-6 gives the possible molecular shapes and examples derived from E.P. shapes.

# of bonded areas	# of un-bonded electron pairs	E.P Shape	Molecular Shape	Example	3-D shape
4	0	Tetrahedral	Tetrahedral	CH_4	
3	1	Tetrahedral	Trigonal pyramid	Ammonia	
3	0	Trigonal Planar	Trigonal planar	H_2CO	
2	2	Tetrahedral	Bent	water	
2	1	Trigonal Planar	Bent	O_3, ozone	
2	0	Linear	linear	CO_2	O=C=O

Comparison of Properties of Ionic and Covalent Compounds

Because of the nature of ionic and covalent bonds, materials produced by these bonds tend to have quite different macroscopic properties. Although the atoms within in a covalently bonded molecule are tightly bound to each other, the individual molecules are generally not strongly attracted to other molecules in a larger sample. On the other hand, the atoms (ions) in ionic materials show strong attractions to other ions in their vicinities. This generally leads to low melting points for covalent solids and high melting points for ionic solids. For example, the molecule carbon tetrachloride is a non-polar covalent molecule (CCl_4) with a melting point of -23°C.

However, the ionic solid NaCl has a melting point of 800°C. Table 5-6 lists some important differences between ionic and covalent compounds.

Ionic compounds	Covalent compounds
1. Crystalline solids (made of ions)	1. Gases, liquids, or solids (made of molecules)
2. High melting and boiling points	2. Low melting and boiling points
3. Conduct electricity when melted or dissolved in water	3. Poor electrical conductors in all phases
4. Most are soluble in water but not in nonpolar liquids	4. Many soluble in nonpolar liquids but not in water

Table 5-6 comparison of ionic and covalent compounds

When salts or ionic compounds dissolve in water they form electrolytes. **Electrolytes** are compounds that conduct electricity when melted or dissolved in water, see Figure 5-13. In your body, electrolytes are minerals found in your blood and other fluids that carry an electrical charge. In the blood, these electrolytes exist as acids, bases, and salts (such as sodium, calcium, potassium, chlorine, magnesium, and bicarbonate).

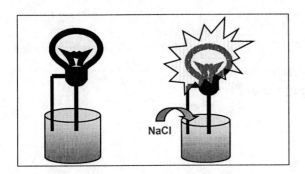

Figure 5-13

Electric current is produced when salt is added to water

Electrolytes are important because they affect the amount of water maintained in your body and its blood pH, muscle action, and other very important processes.

Intermolecular Forces

Where as ionic compounds have strong electrostatic attractions between ions with opposite charges, molecular compounds have an altogether different set of forces acting between them called **intermolecular forces** (attractions between individual molecules). These forces include hydrogen bonding and dipole-dipole; they are responsible for the different properties associated with molecules like boiling points, melting points, viscosity, and density. **Dipole-dipole** interactions are attractive forces between polar molecules. Like polar bonds, molecules can have a charge distribution that results in a dipole. When two polar molecules are near each other the positive end of one will be attracted to the negative end of the other, as shown in Figure 5-14.

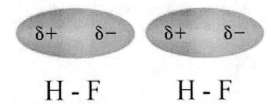

Figure 5-14
Dipole-dipole attraction between to hydrogen fluoride molecules

Hydrogen bonding is a result of the strong interaction between a hydrogen atom bonded to a highly electronegative atom, such as O, N or F, in one molecule and an electronegative atom in another of the same molecule. This inter-molecular force is quite strong and is responsible for the high boiling point associated with water, see Figure 5-15.

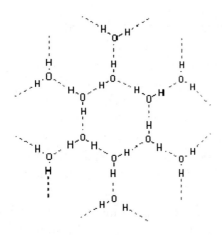

Figure 5-15
Hydrogen bonding in water

Table 5-7 compares the boiling points for hydrogen and non-hydrogen bonded molecules.

HB	BP °C	Non-HB	BP °C	Non-HB	BP °C
H_2O	100	H_2S	-50	H_2Se	-30
HF	20	HCl	-75	HBr	-55
NH_3	-10	PH_3	-80	AsH_3	-70

Figure 5-16 boiling point comparisons for hydrogen bonded and non-hydrogen bonded molecules

States of Matter

We can now see why different substances at the same temperature exist as different states--solid, liquid or gas. **Solids** are defined as having a fixed shape and volume and are non-compressible. The solid state is a result of very strong intermolecular forces acting between molecules. **Liquids** are defined as having a variable shape and fixed volume and are also non-compressible. Although the inter-molecular forces associated with liquids are strong, they are not as strong as those found in solids. **Gases,** on the other hand, are defined as having variable shape and volume and are easily compressed. These properties are associated with weak intermolecular forces between gas molecules, see Figure 5-17.

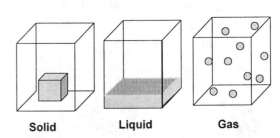

Figure 5-17
States of matter

Chapter 5 Exercises

1. Magnesium nitride is made up of magnesium ions and nitride ions. What is the expected formula of magnesium nitride?

   _____

2. Which is not true about the ionic compound sodium chloride (NaCl)?

   a. it is electrically neutral   b. it was formed when electrons were shared   c. it has properties different from the atoms from which it is formed   d. it is a white crystalline solid

3. A binary ionic compound consists of a

   a. metal and nonmetal   b. metal and metal   c. nonmetal and nonmetal   d. all these possibilities

4. Predict the formula for the compound between calcium and chlorine

   _____

5. Which is expected to form a negative ion?

   a. K   b. Ne   c. Na   d. F

6. How many un-bonded electron pairs are in the water molecule?

   a. 0   b. 2   c. 4   d. 6

79

7. Of the following, which is the strongest bond?

   a. C:C, single bond   b. C::C, double bond   c. C:::C, triple bond   d. all are of equal strength

8. Draw the Lewis structure of ammonia (NH_3), and determine its electron pair shape.

   Electron pair shape _____

9. What is the molecular shape of ammonia?

   a. linear   b. tetrahedral   c. Trigonal pyramid   d. planar (triangular)

10. Draw the Lewis structure of ammonia (CS_2), and determine its electron pair shape

   Electron pair shape _____

11. What is the molecular shape of CS_2?

a. linear   b. tetrahedral   c. trigonal pyramid   d. planar (triangular)

12. List the properties of ionic compounds?

_____

_____

_____

_____

_____

13. The charge of an ion which contains 16 protons, 16 neutrons, and 18 electrons is:

a. 0   b. +2   c. -2   d. -18

14. How many valence (bonding) electrons does the phosphorus atom have?

a. 7   b. 5   c. 15   d. 31

15. What ion is likely to be formed by magnesium (Mg) in chemical reactions?

a. Mg^{1+} b. Mg^{2+} c. Mg^{1-} d. Mg^{2-}

16. Which pair of atoms is most likely to be covalently bonded?

a. Na and O   b. K and F   c. Li and O   d. C and S

17. Which of the following substances has polar covalent bonds? (There could be more than one)

   a. CaF_2 b. KBr   c. NH_3 d. CH_4

18. Which substance has ionic bonds between atoms? (There could be more than one)

   a. Na_2O b. $CaCl_2$ c. H_2S d. CO_2

19. Draw the Lewis structures of both CH_4 and PCl_3 and give their molecular shape

   Molecular shape CH_4_____

   Molecular shape PCl_3_____

20. Which substance is most likely to have covalent bonds between atoms?

   a. $AlCl_3$ b. KBr   c. MgF_2 d. PCl_3

21. A new element has the symbol X and seven valence electrons. What is the maximum number of hydrogen atoms that can bond to X?

a. 1   b. 4   c. 3   d. 7

22. Which of the following pairs of atoms share electrons equally?

a. Cl and H   b. Mg and Cl   c. C and C   d. S and F

23. Define electronegativity

_____
_____
_____
_____
_____

24. An atom has an electronic arrangement of 2-8-8-2. This atom is most likely to

a. form a +2 ion   b. bond covalently to nonmetals   c. form a -2 ion   d. bond covalently to metals

25. A neutral atom can be converted to a negative ion if it

a. gains electrons   b. loses electrons   c. gains protons   d. loses protons

26. What particles make up the crystal in a grain of table salt (sodium chloride)?

   a. atoms of sodium and chlorine   b. molecules of sodium and chlorine

   c. sodium ions and chloride ions   d. molecules of sodium chloride

27. Which of the following has a linear molecular shape?

   a. CH_4 b. SO_3 c. H_2O d. CO_2

28. There are _____ electrons in the covalent bond between C and O in CO.

29. Which of the following molecules has a triple bond?

   a. N_2 b. NH_3 c. C_4 d. H_2O

30. Which of the following will not have hydrogen bonding?

   a. HF   b. NH_3 c. H_2O d. CH_4

31. Particles in the _____ state are in direct contact but are free to move.

32. Define the following by their properties:

Liquids

_____

_____

_____

Solids

_____

_____

_____

Gas

_____

_____

_____

33. Most substances that have molecules with approximately the same weight as water molecules are gases at room temperature. Water is a liquid at room temperature because of

a. extensive hydrogen bonding between water molecules

b. ionic bonding between hydrogen ions and oxide ions

c. covalent bonding between hydrogen atoms and oxygen atoms

d. the lack of motion of water molecules

34. Which of the states of matter has particles in fixed positions and touching one another?

a. solid   b. liquid   c. plasma   d. gas

35. Which of the following molecules has only single bonds?

a. N_2 b. O_2 c. CO_2 d. NH_3

Chapter 6

Chemical Reactions

A **chemical reaction** occurs when two or more chemical substances are mixed together and change into new substances. For this to happen, the bonds between atoms and molecules must break and then re-form in different ways. As we saw in the previous chapter, these chemical bonds can be strong; therefore, energy, usually in the form of heat, is often needed to start a chemical reaction. The **products**, substances formed during the reaction, have different properties from the **reactants**, the original substances that combine in a chemical reaction. The mixing of different substances does not always result in a chemical reaction. Also, some chemical reactions occur so slowly, that they appear not to be reacting at all, while others react violently and can be very harmful if not carefully controlled. Controlling chemical reactions requires an understanding of the factors that influence the rates at which these reactions proceed. For instance, the rusting of iron, a chemical reaction with oxygen, is drastically slowed or even prevented by coating the surface with a chemical rust inhibitor.

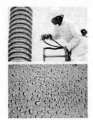

Figure 6-1
 Rust inhibitor being applied to iron coils

It is sometime beneficial for reactions speeds to be increased, for example, the faster a medicinal drug can enter the blood-stream, the faster it can do its job of reducing a fever, easing a pain, or even clearing a stuffy nose.

In this chapter, we will discuss the details of chemical reactions and answer some key questions such as:

What are balanced equations?

What are moles?

What are reaction rates, and how do we influence them?

What are equilibrium reactions?

What are the first and second laws of thermodynamics?

The Chemical Equation

A **chemical equation** is a shorthand molecular description of the identities and amounts of the reactants and products of a chemical reaction. For example, silver (Ag) reacts with sulfur (S) in the air to make silver sulfide, (Ag_2S) the black material we call tarnish.

$$2\ Ag\ {(s)} + S\ {(g)} \rightarrow Ag_2S\ {(s)}$$

According to this equation, two molecules of silver react with one molecule of sulfur forming (indicated by the →) one molecule of silver sulfide (tarnish). Also, methane (CH_4), sometimes called natural gas, reacts with oxygen in the air forming carbon dioxide and water:

$$CH_4\ {(g)} + 2\ O_2\ {(g)} \rightarrow CO_2\ {(g)} + 2\ H_2O\ {(l)}$$

This equation tells us that one molecule of methane will react with two molecules of oxygen, forming one molecule of carbon dioxide and two molecules of water.

Very often, as in the equations above, you will see the descriptions of the materials in the reaction in parentheses after the material. A gas is shown by (*g*), a solid by (*s*), and a liquid material by (*l*). A material dissolved in water (an aqueous solution) is shown by (*aq*). An upwards arrow (↑) indicates a gas being formed, whereas a downwards arrow (↓) indicates a solid precipitate being formed.

The equations shown above are also said to be balanced if they follow Dalton's atomic theory- that matter is conserved during the course of a chemical reaction (Chapter 2). This means that the number and kinds of atoms on the left side of the reaction arrow (reactants) must be the same as that on the right side of the arrow (products). Look at the combustion of methane again:

$$CH_4 \text{ (g)} + 2\ O_2 \text{ (g)} \rightarrow CO_2 \text{ (g)} + 2\ H_2O \text{ (l)}$$

Notice that on the left side of the arrow there is one atom of carbon, four atoms of hydrogen, and four atoms of oxygen. On the right side there is one atom of carbon, four atoms of hydrogen and four atoms of oxygen. This equation is balanced because the same number and kinds of atoms are present on both sides of the reaction arrow (1 C atom, 4 H atoms and 4 O atoms). Now is a good time to talk about the difference between subscripts and coefficients. When a single molecule exists as a combination of different atoms, **subscripts** are used to show the numbers of different kinds of

atoms present. For example, methane (CH_4) is a single molecule that contains one carbon and four hydrogen atoms. On the other hand, coefficients are used to represent the number of molecules present. For example, if two molecules of methane were present in an equation, it would be written as $(2CH_4)$. This tells us there are now two carbon atoms and eight hydrogen atoms (four on each methane molecule) present.

Example 6-1

Determine the numbers of each atom present in the following molecule.

a) barium hydroxide; $Ba(OH)_2$

b) ammonium sulfide; $(NH_4)_2S$

Answer

a) The subscript 2 on the outside of the parentheses tells us there are two OH entities (OH, OH); therefore, there are two oxygen atoms and two hydrogen atoms present and, of course, only one barium atom.

B atoms = 1
O atoms = 2
H atoms = 2

b) Here, the subscript on the outside of the parentheses tells us there are two NH_4 entities (NH_4, NH_4), therefore, there are two nitrogen and eight hydrogen atoms and, of course, only one sulfur atom.

S atoms = 1
N atoms = 2
H atoms = 8

88

Example 6-2

Is the following equation is balanced?

$$Ca(OH)_2 + 2\,HCl \rightarrow CaCl_2 + H_2O$$

Answer

The number and kinds of atoms on each side of the arrow are:

Left side   Right side

Ca = 1   Ca = 1
Cl = 2   Cl = 2
O = 2   O = 1
H = 4   H = 1

No, the equation is not balanced. Although there are the same numbers of calcium and chlorine atoms on both sides, the numbers of oxygen and hydrogen atoms are different.

Balancing Chemical Equations

In general there are four steps involved in balancing chemical equations:

1. Write all reactants on the left and all products on the right side of the equation arrow. Make sure you write the correct formula for each element.
2. Use coefficients in front of each formula to balance the number of atoms on each side.
3. Multiply the coefficient of each element by the subscript of the element to count the atoms. Then list the number of atoms of each element on each side.

4. It is often easiest to start balancing with an element that appears only once on each side of the arrow. These elements must have the same coefficient. Next, balance elements that appear only once on each side but have different numbers of atoms. Finally, balance elements that are in two formulas in the same side.

Example 6-2

Balance the following equation:

$$NH_3 + O_2 \rightarrow NO + H_2O$$

Answer

N appears once on both sides in equal numbers, so the coefficient for NH_3 is the same as for NO.

Next, look at H, which appears only once on each side but has different numbers of atoms (3 on the left and 2 on the right). The least common multiple of 3 and 2 is 6, so rewrite the equation to get 6 atoms of H on both sides:

$$2NH_3 + O_2 \rightarrow NO + 3H_2O$$

The number of nitrogen atoms on the left is now different than that on the right. Rewriting the equation to get the same number of nitrogen atoms on both sides gives:

$$2NH_3 + O_2 \rightarrow 2NO + 3H_2O$$

There are now 2 oxygen atoms on the left and 5 on the right. The least common multiple of 2 and 5 is 10, so rewriting the equation to give 10 oxygen atoms on both sides gives:

$2NH_3 + 5O_2 \rightarrow 4NO + 6H_2O$

Now count the atoms on each side:

Left side	Right side
2 N	4 N
6 H	12 H
10 O	10 O

If you double the N and H on the left, the equation will be balanced:
$4NH_3 + 5O_2 \rightarrow 4NO + 6H_2O$
Double-check: by counting the atoms

Left side	Right side
4 N	4 N
12 H	12 H
10 O	10 O

Example 6-2

Balance the following equation:

$$C_5H_{12} + O_2 \rightarrow CO_2 + H_2O$$

Answer

In this case there are no atoms that appear only once and in equal number. Carbon appears only once but has different numbers--5 on the left and 1 on the right. Rewriting the equation to balance the carbons gives:

$$C_5H_{12} + O_2 \rightarrow 5CO_2 + H_2O$$

Next look at H which appears only once on each side but also has different numbers of atoms--12 on the left and 2 on the right. So rewriting the equation to get 12 on the left side gives:

$$C_5H_{12} + O_2 \rightarrow 5CO_2 + 6H_2O$$

There are now 2 oxygen atoms on the left and 16 on the right. A coefficient of 8 in front of O_2 on the left will give 16 oxygen's and gives:

$$\mathbf{C_5H_{12} + 8O_2 \rightarrow 5CO_2 + 6H_2O}$$

Now count the atoms to see if it is balanced.

Left side	Right side
5 C	5 C
12 H	12 H
16 O	16 O

It is balanced.

What does a balanced equation tell us?

Balanced equations give the stoichiometric relationships of chemical reactions. **Stoichiometry** deals with the quantities of substances that enter into, and are produced by, chemical reactions. Although this word may sound complicated, the idea of stoichiometry is not. If we first apply stoichiometry to relationships found in areas other than chemistry, we will see it to be quite logical and simple. Let's say, for instance, that you build tables that are made from four legs and one table top. The balanced equation or relationship for the building of tables could be given as:

$$4 L + 1TT \rightarrow 1TB$$

Where L = legs, TT = table tops, and TB = table.
Obviously, if you had exactly 4 legs and 1 table top you could build exactly 1 table. So, how many tables could you build if you had 36 legs and seven table

90

tops? You could answer this in two ways. First, 7 table tops requires 28 legs to build 7 tables, which is the maximum number you could build based on the number of tops. Second, 36 legs would make 9 tables (with four legs 4 per table). Therefore, you have enough legs to make 9 tables but only enough table tops to make 7 tables. The maximum amount of tables you could build is, of course, 7 even though you have 8 legs remaining.

This is what stoichiometry is all about, and it can be applied to relationships found in chemical reactions. For example, look at the following balanced reaction involving the combustion of pentane (C_6H_{12}) with oxygen and the production of carbon dioxide and water:

$$\boxed{C_5H_{12} + 8O_2 \rightarrow 5CO_2 + 6H_2O}$$

There are many relationships in this equation. For instance, one molecule of pentane will produce five molecules of carbon dioxide if enough oxygen is available. Also, if you wanted to produce twelve molecules of water you would need to burn two molecules of pentane in the presence of excess oxygen.

Example 6-3

Use the balanced equation below to answer the following questions.

$CH_4 + 2O_2 \rightarrow CO_2 + 2H_2O$

a) How many molecules of water could form by burning five molecules of methane (CH_4) with excess oxygen?

b) How many molecules of methane would you have to burn in excess

---

oxygen to produce 12 molecules of water?

Answer

a) The balanced equation tells us that one molecule of methane will produce two molecules of water when burned in excess oxygen. In other words, the ratio of oxygen to methane is 2:1.

$$\frac{2\,H_2O\ \text{molecules}}{1\,CH_4\ \text{molecules}}$$

Using 5 molecules of CH_4 We can set up the ratio:

$$\frac{2\,H_2O\ \text{molecules}}{1\,CH_4\ \text{molecules}} = \frac{?\,H_2O\ \text{molecules}}{5\,CH_4\ \text{molecules}}$$

Therefore,

$$5\,CH_4\ \text{molecules}\,x\,\frac{2\,H_2O\ \text{molecules}}{1\,CH_4\ \text{molecules}}$$

= 10 H_2O molecules would be formed

b) Like in a) use the following ratio:

$$\frac{1\,CH_4\ \text{molecules}}{2\,H_2O\ \text{molecules}} = \frac{?\,CH_4\ \text{molecules}}{12\,H_2O\ \text{molecules}}$$

Therefore,

$$12\,H_2O\ \text{molecules}\,x\,\frac{1\,CH_4\ \text{molecules}}{2\,H_2O\ \text{molecules}}$$

= 6 CH_4 molecules would be needed

The Mole

Methane, CH₄, burns in oxygen to produce carbon dioxide and water.

$$CH_4 + 2O_2 \rightarrow CO_2 + 2H_2O$$

This equation states that one molecule of CH₄ combines with two molecules of O₂ and produces one molecule of CO₂ and two molecules of H₂O. While this statement is certainly true, the small sizes and masses of these molecules make it impossible to observe the reaction on this scale. For instance, one molecule of water would only weigh 3.0×10^{-26} kg. To put this into perspective, one teaspoon of water contains 1.7×10^{23} molecules of water. This means that there are more molecules of water in one teaspoon of water than there are teaspoons of water in all the oceans of the world. As you can see, molecules are very small. So, how do we measure quantities of molecules so that their ratios are consistent with the ratios found in balanced chemical equations? Chemists count molecules by the mole. A **mole** is simply a number equaling 6.02×10^{23} and is defined as the number of carbon atoms in exactly 12.0 grams of the carbon-12 isotope. This number is quite large and deserves to be put into some perspective. Did you know that a stack of paper with 6.02×10^{23} sheets would be so tall that it would reach the sun--not just once but more than a million times; or if 6.02×10^{23} dollars were divided among all 6 billion people on Earth, we would have enough money that each of us could spend one million dollars every minute for the rest of our lives and still have more than half remaining; or if you could travel at the speed of light, it would take you more than a 100 billion years to travel 6.02×10^{23} miles? Like was said earlier, the mole is simply a number, a very large one, but is used to count molecules the same way the term *dozen* is used to count eggs. For example, a mole of carbon atoms is equal in number to a mole of hydrogen atoms. Furthermore, one mole of methane (CH₄) contains four moles of hydrogen. Similarly, if you had a mole of cars you would have four moles of tires.

Example 6-4

How many moles of each element are present in one mole of the following compound?

N_2O_4

Answer

1 mole of N₂O₄ contains 2 moles of N and 4 moles of O.

Molar Mass of Molecular weight

Recall from Chapter 2 that each element has an associated atomic weight, measured in μ, located underneath the symbol of each element on the periodic table (Figure 6-2).

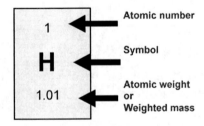

Figure 6-2
Atomic weight of hydrogen found on the periodic table

The atomic weight of an element is also numerically equal to the molar mass of that element. Molar mass is equal to the mass in grams of one mole of an element. For instance, the molar mass of hydrogen is equal to 1.01 grams/mole. This means that 1.0 mole of hydrogen is equal to 1.01 grams of hydrogen. By referring to the periodic table we can see that 22.99 grams of sodium (Na) is equal to one mole of sodium (Figure 6-3).

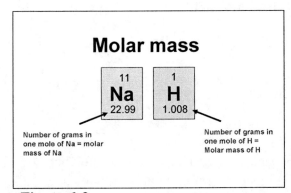

Figure 6-3
Molar mass relationships and the periodic table

Example 6-5

How many grams are in 3.0 moles of aluminum (Al)?

Answer

By referring to the periodic table we see that the molar mass of aluminum is 26.98 g/mol. This means that one mole of aluminum weighs exactly 26.99 grams and, therefore, 3 moles of aluminum would weigh

3 moles x 26.99 grams/mole

= 80.97 grams of aluminum.

Example 6-6

How many moles are there in 65.0 grams of magnesium (Mg)?

Answer

By referring to the periodic table we see that the molar mass of magnesium is 24.31 g/mol. This means that every 24.31 grams of magnesium is equal to one mole of magnesium. Therefore

$$\frac{65.0\,g}{24.31\,g/mol} = 2.67\,mol\;Mg$$

The previous examples show the procedure for converting from moles to grams and grams to moles of a substance. Figure 6-4 summarizes the methods for these conversions.

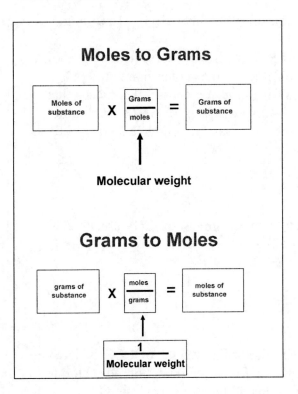

Figure 6-5
Method for converting between moles and grams

Molar Masses of Compounds

Since molecular compounds are combinations of atoms, their molecular weights are determined from the weights of each atom in the molecule. The molecular weight is the sum of the atomic weights of all the atoms present in the molecular formula, expressed in amu. The molar mass of the compounds is numerically the same, but expressed in grams/mole. For instance, the molecular weight of water (H_2O) can be calculated from the sum of the atomic weights for the elements hydrogen and oxygen.

$$1 \text{ atom of O} \left(\frac{15.99 \text{ amu}}{1 \text{ atom O}} \right) = 15.99 \text{ amu}$$

$$2 \text{ atom of H} \left(\frac{1.01 \text{ amu}}{1 \text{ atom H}} \right) = 2.02 \text{ amu}$$

Molecular weight of H_2O = 18.01 amu

Since H_2O has a molecular weight of 18.01 amu, the molar mass of H_2O is 18.01 g/mol. Of course, this means that one mole of water, or 6.02×10^{23} molecules of water, weighs exactly 18.01 grams.

Example 6-6

How many grams of CO_2 are in 2.30 moles of CO_2?

Answer

The molecular weight of CO2 is determined from the sum of the atomic weights of the elements C and O:

$$1 \text{ C atom} \left(\frac{12.01 \text{ amu}}{1 \text{ atom C}} \right) = 12.01 \text{ amu}$$

$$2 \text{ atom O} \left(\frac{15.99 \text{ amu}}{1 \text{ atom O}} \right) = 31.98 \text{ amu}$$

Molecular weight of CO_2 = 43.99 amu

Therefore the molar mass of CO_2 is

43.99g/mol

Meaning exactly one mole CO_2 weighs exactly 43.99 grams

Therefore in three moles:

$$3 \text{ mol } CO_2 \left(\frac{43.99 \text{ g } CO_2}{1 \text{ mol } CO_2} \right) = 131.97 \text{ g } CO_2$$

You can also convert from the number of grams to moles of a compound.

Example 6-7

How many moles of CO_2 are present in 55.0 grams of CO_2?

Answer

From Example 6-6 the molar mass of CO_2 is 43.99 g/mol

Therefore:

$$55.0 \text{ g } CO_2 \left(\frac{1 \text{ mole } CO_2}{43.99 \text{ gl } CO_2} \right) = 1.25 \text{mol } CO_2$$

Rates of Reactions

Every chemical reaction occurs, under a given set of conditions, at a characteristic speed called *rate*. The **rate of a reaction** is the speed at which a reaction happens. This means that if a reaction has a low rate, the molecules combine at a slower speed than another with a high rate. Some reactions take hundreds, maybe even thousands, of years while others can happen in less than a second. Reaction rates are very dependent upon factors such as the concentration, temperature, and pressure of the reactants as well as the presence or absence of a catalyst. **Kinetics,** the study of reaction rates, gives scientists an insight into a variety of aspects, one of them being what can be done to increase or decrease the rate of a chemical reaction. The knowledge of reaction kinetics has many practical applications such as in designing therapeutic drugs that metabolize faster and thereby enter the blood stream sooner.

There are quite literally billions and billions of chemical reactions occurring in the human body every minute, proceeding at very different rates. For example, red blood cells are replaced at a rate of about 100 billion a day, while nerve and brain cells are replaced very slowly if replaced at all.

What has to happen for a reaction to occur?

For any reaction to occur, three things must happen. First, the reactants must come in contact with each other, secondly, they must collide with enough energy to overcome the activation energy; last, they must have proper orientation when they collide. The first criterion for the occurrence of a reaction is quite obvious, because a reaction cannot occur if the reactants are in separate containers. The second however may be less obvious. **Activation energy** is defined as the minimum amount of energy required for a reaction to occur. As the reacting molecules approach each other, their valence electrons repel each other. Overcoming this repulsion requires energy (activation energy; see Figure 6-6), which is usually provided in the form of heat. If there is enough energy available, the repulsion is overcome and the molecules can get close enough for a reaction to occur.

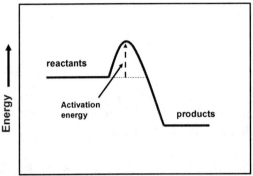

Figure 6-6
    Energy level diagram. The activation energy is represented by the hump in the diagram

As the heat or temperature of the reaction is increased, more molecules collide with sufficient energy to overcome the activation energy, and as a result, the reaction rate increases. The third criterion requires that the colliding molecules have a proper orientation. Molecules that collide with sufficient energy, overcoming the activation energy but not properly oriented, will simply bounce off one another and not form products (Figure 6-7).

Controlling Reaction Rates

As the collision frequency increases, the number of molecules colliding with proper orientation and energy also increases and, therefore, increases the rate of the reaction. One way to cause this to happen is to increase the temperature of a reaction, which will increase the average speed of the molecules, allow them to collide more frequently and with greater energy and, therefore, increase the rate.

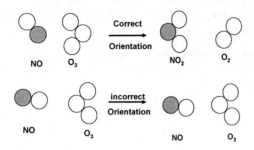

Figure 6-7
Not every collision results in the formation of products, due to improper orientation of colliding molecules

On the other hand, lowering the temperature will have the opposite effect and decrease the reaction rate.

Another way to increase the frequency of collisions is to increase the number of reacting molecules in a given volume. For reactions in a solution, adding more of the reactants or reducing the amount of solvent will provide for a higher frequency of collisions. For gases, increasing the pressure or decreasing the volume will in increase the concentration and, again, increase the collision frequency (Figure 6-8).

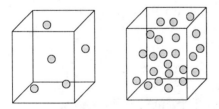

Figure 6-8
Increasing concentration increases the number of collisions per second.

Another method for increasing the number of **productive collisions,** the number of molecules colliding with proper orientation and energy to affect a reaction, is to lower the activation energy. A **catalyst** is a substance that increases the rate of reaction but is not consumed in the reaction. Catalysts provide an alternate path, with a lower activation energy, for molecules to follow in their conversion to products. Figure 6-9 shows an energy diagram for a catalyzed and an un-catalyzed reaction.

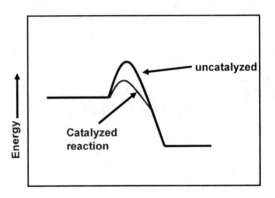

Figure 6-9
Energy diagram for a catalyzed and un-catalyzed reaction

Enzymes are catalysts found in biological systems. For example, the enzyme carbonic anhydrase, found in red blood cells, increases the rate of formation of the bicarbonate ion, from carbon dioxide and a hydroxide ion, by a factor 3,000,000. Without this enzyme, the reaction would be so slow that the body could not rid itself of the carbon dioxide generated by cellular processes.

Reactions at Equilibrium

So far we have looked at reactions proceeding in one direction only--reactants forming products.

Reactants $\rightarrow$ Products

Reactants are shown on the left side of the arrow and the products on the right side. However, most chemical reactions can proceed in both directions, allowing products to form reactants.

Reactants ← Products

When the rate of the forward reaction is equal to the reverse reaction, the system is said to be in a state of **equilibrium**. For example, icebergs are in a state of physical equilibrium, in that they are freezing and melting at the same rate and appear not to change over time.

Figure 6-10
Icebergs are melting and freezing at the same rate and appear not to change over time, so they are said to be in a state of physical equilibrium.

Physical equilibrium exists when the rates at which a substance changes between physical states (i.e. solid, liquid, and gas) are equal. Equilibrium reactions are written like other reactions with the exception of a double arrow pointing in both directions. For example, the physical equilibrium that exists for the melting and freezing of icebergs is written as:

$$H_2O\ (s) \leftrightarrow H_2O\ (l)$$

Chemical equilibrium exists when the rates of the forward and reverse reactions in a *chemical* change are equal.

For example, when a sample of nitrogen dioxide (NO_2), a brown gas, is placed in a sealed container, it will decompose and form dinitrogen tetroxide (N_2O_4), a colorless gas; this gas will, in turn, decompose and reform nitrogen dioxide. During this process, the color of the gas becomes lighter as N_2O_4 is produced and continues until no further reaction is observed. At this point, the reaction is at equilibrium. Meaning the rate at which the NO_2 is forming N_2O_4 is equal to the rate at which N_2O_4 is forming NO_2.

$$2NO_2\ (g) \leftrightarrow N_2O_4\ (g)$$

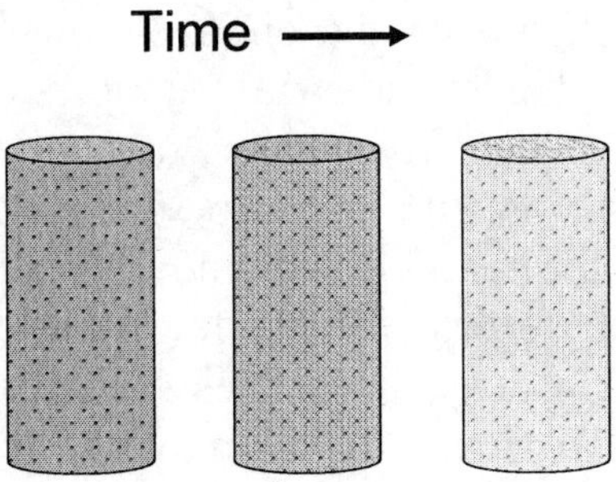

Time ⟶

$$2NO_2 \rightleftharpoons N_2O_4$$

Figure 6-11
Equilibrium mixture of NO_2 and N_2O_4

When a system at equilibrium is disrupted, the system will move in a direction to reestablish equilibrium. This is **Le Chatelier's Principle**. For example, in the reaction shown in Figure 6-11, the concentrations of NO_2 and N_2O_4 in the container on the right, are constant in time. If, however, you removed some of the N_2O_4 from the container, you will have disrupted the equilibrium. According to Le Chatelier's principle, the reaction will then proceed to the right (forming more N_2O_4) in order to reestablish equilibrium. If the reaction involves only gases, an increase in pressure to the system will cause the reaction to proceed in the direction that contains the fewer number of moles of gas. This is because fewer numbers of moles of gas require less space and can handle the increased pressure. Likewise, if the pressure is decreased, the reaction will proceed in the direction with the greater number of moles. For example, the equation representing the equilibrium reaction of nitrogen with hydrogen, forming ammonia (NH₃)

$$3H_2(g) + N_2(g) \leftrightarrow 2NH_3(g)$$

shows four total moles of gas (three moles of hydrogen and one mole of nitrogen) on the left side and two moles of gas on the right (two moles of ammonia). If the pressure of this system is increased, the reaction will move to the right because that side has fewer moles of gas. However, if the pressure is decreased, the reaction will move to the left. Also, according to Le Chatelier's principle removing or adding ammonia to the system will shift the reaction in the direction where it can make up for the lost or gained ammonia (Figures 6-12 and 6-13).

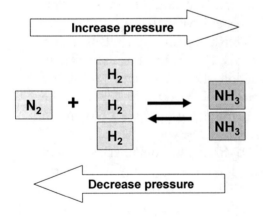

Figure 6-12

An increase in pressure moves the reaction to the right and a decrease moves the reaction to the left

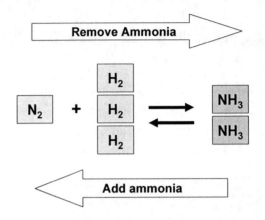

Figure 6-13
Removing ammonia moves the reaction to the right and adding ammonia shifts the reaction to the left.

What are the Driving Forces for Chemical Reactions?

Driving forces are those factors which are responsible for causing a reaction to occur. All chemical processes need to have some reason for taking place. There are two main requirements which seem to drive all chemical reactions[5]. These requirements include a release of energy and an increase in entropy. Chemical reactions that result in a release of energy (usually in the form of heat) are called **exothermic reactions,** and those that result in the gain of energy are called **endothermic reactions.** Since all systems strive to release energy, all chemical processes would prefer to undergo an exothermic process. On the other hand, **entropy** is a measure of the

---

[5] For the purpose of this discussion we will us the term "system" to describe a chemical reaction.

disorder of a system--the higher the disorder, the larger the entropy. And since all systems strive to increase their entropies, all chemical processes would prefer to undergo a change that leads to a state of greater disorder. Examples of changes that lead to a more disordered state include the melting of ice, which forms a liquid, and a liquid evaporating, which forms a gas. Changes in entropy are related to changes in the randomness of the system. Most changes in entropy can by predicted from a few general concepts:

1)   The entropy of a substance increases when a solid forms a liquid and a liquid forms a gas
2)   Entropy generally increases when a solid dissolves in a liquid.
3)   Entropy decreases when a gas dissolves in a liquid.
4)   Entropy increases as the temperature increases.

Example 6-8

Determine if the following processes give an increase or decrease in entropy

a) Butter is melted

b) Rain turns into snow

c) A new deck of cards is shuffled

Answer

a) Solid butter is more ordered than liquid butter, therefore, the entropy increases

b) Snow is more ordered that liquid water, therefore, entropy decreases

---

c) A new deck of cards is usually arranged by numbers and suits; shuffling disrupts this order and, therefore, the entropy increases.

Although both a decrease in energy and an increase in entropy are responsible for chemical reactions to occur spontaneously, both do not have to be present for a spontaneous reaction to take place.

A **spontaneous reaction** is a process in which at least one of the driving-force requirements has been met, allowing the reaction to proceed without any outside influences. In such a system, an additional factor exists for the process to take place. Some of these processes will require a specific temperature in order to occur, while others will occur at any temperature. The reaction, however, may not happen automatically. It may be necessary to input energy, or use some other mechanism to get it started, but it will ultimately occur. For instance, the burning of gasoline, a process that requires a "spark" to get started, will spontaneously burn to completion. In this example, the exothermic requirement was met, however, the reaction resulted in a decrease of entropy. This can be seen from looking at the balanced chemical equation for the combustion of gasoline:

$$2C_8H_{18}\ (l) + 25\ O_2\ (g) \rightarrow$$
$$16CO_2\ (g) + 18H_2O\ (l)$$

Also, as was mentioned earlier, entropy increases when a solid changes to a liquid and a liquid changes to a gas. In the combustion of gasoline we can see that there are twenty-five moles of gas on the reactant side and only sixteen moles of gas on the product side. However, there are more moles of liquid on the product side compared to the reactant side. From

99

these observations we can conclude that the system has definitely become more ordered, resulting in a decrease in entropy.

Energy requirement Satisfied	Entropy requirement satisfied	Reaction temperature
Yes	Yes	All Temperatures
Yes	No	Low Temperature
No	Yes	High Temperature
No	No	Not Spontaneous at any Temperature

Table 6-1
        Conditions for a spontaneous change to occur

It can then be surmised that at least one of the driving forces must be present for a chemical reaction to proceed spontaneously. However, if only one driving force is present, additional factors, such as specific reaction temperatures, may be required.   Table 6-1 summarizes the conditions for a spontaneous change to occur.

From Table 6-1 we can see that any chemical reaction will occur spontaneously, at any temperature, if the reaction releases energy (exothermic) and the entropy increases, becoming more disordered. If the reaction is endothermic and the entropy decreases, then neither requirement is satisfied and the reaction will not occur spontaneously at any temperature.

Laws of Thermodynamics

Thermodynamics is the study of the conversion of energy from one form to another. For example, burning gasoline converts the chemical energy (in the gas) to heat, which can then be used as mechanical energy to power your car. Also, the kinetic energy of flowing water that is used to turn a hydroelectric generator is converted into electrical energy. The conditions for any change in energy are summarized by the first and second laws of thermodynamics. The **first law of thermodynamics**, often called the **law of conservation of energy**, suggests that energy can neither be created nor destroyed only transferred from one form to another. To put it another way, the total amount of energy available in the universe is constant.

Einstein's famous equation ($E=mc^2$) describes the relationship between energy and matter. This equation states that energy (E) is equal to mass (m) times the square of a constant (c). This equation suggests that matter and energy are interchangeable and that there is a fixed quantity of matter and energy in the universe.

The **second law of thermodynamics** addresses the entropy change in any process. This law suggests that the total entropy of the universe is constantly increasing. As mentioned earlier, increasing the temperature of an object increases its entropy. This is because hot molecules move at a more random and faster rate. It's also due to the fact that energy (i.e. heat) transfer can only occur in one direction; the idea that heat always moves from hot to cold is the basis for this law. The consequence of this law is that once energy is converted to entropy it is no longer available for a useful purpose. For example, once the energy stored in the chemical bonds of gasoline is released during combustion, the energy in the form or heat is transferred to the surroundings, increasing the entropy of the universe, and will no longer be available to do useful work.

Useful Chemical Reactions

The three types of reactions that are most beneficial to people of industrialized countries are combustion, oxidation--reduction, and neutralization. It is, therefore, a good idea to examine these reactions and discuss their useful applications.

Combustion Reactions

A combustion reaction occurs when all substances in a compound combine with oxygen. Combustion is commonly called burning. These reactions are almost always exothermic

Figure 6-14
Combustion of a match

in that they release heat. Gasoline, oil, and wood are examples of *organic materials* or *compounds* (Chapter 8). Organic compounds are made from carbon, hydrogen and oxygen only, and when they burn they produce carbon dioxide, water, and a lot of heat. For example, consider the combustion of methanol (rubbing alcohol):

CH_3OH (l) + O_2 (g) → CO_2 (g) + H_2O (l) + Heat

However, many combustion reactions occur with compounds or elements that are not organic, like for instance the reaction of magnesium and oxygen:

$2Mg$ (s) + O_2 (g) → $2MgO$ (s) + Heat

Notice that carbon dioxide and water are not produced in the combustion of magnesium. These products are limited to combustions of organic materials.

Applications of combustion reactions include the obvious, like fuel sources for cars, boats, planes, motorcycles, lawnmowers etc. Also, the combustion of natural gas and coal is used to heat and provide electricity for many homes and businesses.

Oxidation Reduction Reactions

The term "oxidation" originated from the process of combining oxygen with other elements, forming oxides. And the term "reduction" stems from the removal of oxygen from an oxide. However, the combining of elements with chlorine, bromine, and other non-metals were noticed to be very similar to those involving oxygen and, therefore, the definitions were expanded and are given as:

Oxidation- the gain of oxygen, the loss of hydrogen, or the loss of electrons

Reduction- the loss of oxygen, the gain of hydrogen, or the gain of electrons

These definitions are summarized in Table 6-2

Oxidation	Reduction
Loss of an electron	Addition of an electron
Addition of oxygen	Loss of oxygen
Loss of hydrogen	Addition of hydrogen

Table 6-2

Definitions of oxidation and reduction

Although there are three separate definitions for both oxidation and reduction, the last, loss or gain of electrons, encompasses the others. In order to clearly see this connection we need to consider and define oxidation numbers. **Oxidation numbers**, sometimes called **oxidation states**, are either positive or negative integers assigned to an element based on the number of electrons lost or gained. For example, the reaction of sodium with chlorine, producing sodium chloride results in a net loss of one electron for sodium and a net gain of one electron for chlorine.

$$2Na_{(s)} + Cl_{2(g)} \longrightarrow 2NaCl_{(s)}$$

Recall, from Chapter 3, that ionic bonds are formed from the transfer of electrons from one element to another. Also, the direction of this transfer occurs from the element with a lower to the element with a higher electronegativity. Sodium has a much lower electronegativity and, therefore transfers its electron to chlorine producing the sodium ion (Na^+) and the chloride ion (Cl^-). This results in the oxidation of sodium (loss of an electron) and the reduction of chlorine (gain of an electron). The oxidation number of sodium is, therefore, (+1) and that of chlorine is (-1). From this it can be stated that the oxidation numbers for monatomic (one atom) ions is equal to the charge on the ion. For covalent compounds the oxidation numbers are not so easily determined. The differences in electronegativities, however, are very important in assigning their oxidation numbers. Take, for example the reaction of hydrogen and oxygen forming water:

$$2H_{2\,(g)} + O_{2\,(g)} \rightarrow 2H_2O_{(l)}$$

Both reactants, H_2 and O_2, are uncombined neutral elements and are therefore assigned an oxidation state, or number, of zero. The hydrogen-oxygen covalent bond in the water molecule is formed from the sharing of two electrons; one from hydrogen and one from oxygen. Because oxygen is more electronegative than hydrogen, but electrons are not shared equally; the electrons are associated more with the oxygen atom. This can be viewed as a net gain of one electron for the oxygen atom and a net loss of one electron for the hydrogen atom. Since there are two hydrogen atoms covalently bonded to one oxygen atom in a water molecule the net number of electrons gained is 2. We can then assign an oxidation number of -2 to oxygen and +1 for each hydrogen atom and report oxygen as being reduced and hydrogen as oxidized. The importance of assigning oxidation states to elements in balanced chemical equations is that it helps us determine if an oxidation or reduction has occurred. Furthermore, oxidation and reduction always occur together. For instance, if it is determined that an oxidation has occurred in a chemical re-action then a reduction must have also occurred. Atoms can only lose electrons when there is something there to accept them. Since these reactions always occur together, they are often called **redox** re-actions.

In determining if a chemical equation represents a redox reaction the assigning of oxidation states for each element is necessary. If an elements ox-idation state changes in the course of the reaction then redox has occurred. As-signing oxidation states for elements in chemical equations is easily accom-plished by following the set of guide-lines outlined below.

Rules for assigning oxidation states

1. Oxidation states for elements in an uncombined form is zero

2. The oxidation state of a monatomic ion is the charge on the ion

3. In compounds, F is always –1 and the other halogens are –1 unless combined with oxygen or halogen above it, then it is +1.
4. In compounds, H is +1.

5. In compounds, O is always –2 unless combined with F, then it is +2

6. The sum of the oxidation states must be zero for a neutral

---

compound or equal to the charge on a polyatomic ion.

Example 6-9

Is the following a redox reaction, and if so, what is being oxidized and what is being reduced?

$$CH_4 \text{ (g)} + 2O_2 \text{ (g)} \rightarrow CO_2 \text{ (g)} + 2H_2O \text{ (}l\text{)}$$

Answer

In order to determine if this is a redox reaction, we have to assign oxidation numbers to each element. This is easiest if we first separate the elements by drawing lines between them:

$$|C|H_4|\text{(g)} + 2|O_2|\text{(g)} \longrightarrow |C|O_2|\text{(g)} + 2|H_2|O|\text{(l)}$$

Using the rules assign oxidation num-bers to the elements :

103

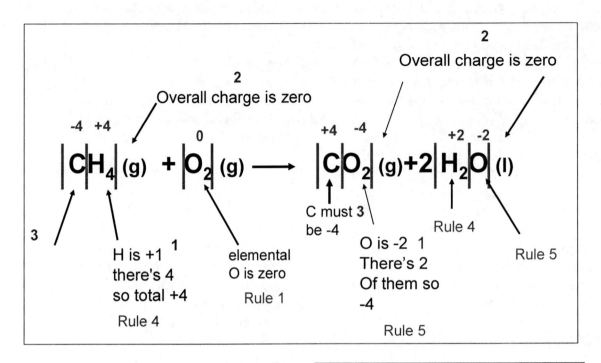

From this we can see the oxidation number of the carbon atom has changed from -4 in CH_4 to +4 in CO_2. And in oxygen it has changed from 0 in O_2 to -2 in both CO_2 and H_2O. Therefore, we can definitely say this is a redox reaction, that carbon was oxidized (losing eight electrons) and all 4 oxygen atoms were reduced (gaining 2 electrons each for a total of 8)

Example 6-10

Assign oxidation numbers for each atom in the polyatomic ion sulfate.

$$SO_4^{2-}$$

Answer

Like in Example 6-9, separate each element with a line and assign oxidation numbers using the rules.

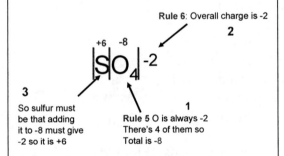

Each oxygen atom has an oxidation number of -2, giving a total of -8, therefore, sulfur must be +6 so that the sum of the oxidation numbers equals the charge on the polyatomic ion -2.

An **oxidizing agent** causes an oxidation of another reactant by accepting electrons from it, and as a result, it becomes reduced. On the other hand, a **reducing**

agent causes the reduction of another reactant by giving it electrons, and as a result, it becomes oxidized. Oxidizing and reducing agents are used in many industrial processes and are found in many consumer products such as car batteries and household bleach. Table 6-3 and 6-4 lists some common oxidizing and reducing agents and their uses.

Name	Formula	Uses
Oxygen	O_2	Metabolism of foods, and combustion
Lead dioxide	PbO_2	Automobile batteries
Sodium Hypochlorite	NaOCl (aq)	Laundry Bleach disinfectant
Chlorine	Cl_2	Purification of water

Table 6-3
Oxidation agents and their uses

Name	Formula	Uses
Hydrogen	H_2	Fuel, chemical synthesis
Sulfur dioxide	SO_2	Chemical synthesis
Carbon	C	Iron Production
Zinc	Zn	Batteries

Table 6-4
Reducing agents and their uses

Free radicals, a class of oxidizing agents, are very reactive atoms or molecules that contain an unpaired electron. When free radicals come into contact with other atoms or molecules, it will rapidly remove one of their electrons. If the electron is removed from a vital molecule like DNA, its structure will become altered. This can have an adverse effect on the DNA molecules ability to perform its proper functions. Free radicals have been linked to aging, cancer, and other degenerative conditions. Free radicals arise in the body from many different sources such as normal biochemical reactions, as well as from inhalation of toxic substances found in cigarette smoke and pollutants (Figure 6-15).

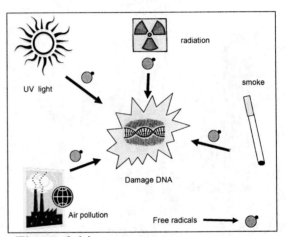

Figure 6-14
Formation of free radicals

These free radicals are quickly deactivated and converted into a less reactive molecule by picking up an electron. **Antioxidants** are substance that inhibits oxidation or reactions promoted by free radicals. The presence of antioxidants in the body helps in the prevention of free radicals combining with and damaging biologically important molecules. Antioxidants are found in many vitamins such as E and C, as well as in many foods, including artichokes, cranberries, blueberries, pecans, and even cinnamon.

Electrochemical Cells and Batteries

Redox reactions in which electrons are transferred from one substance to another substance can be used to produce electricity. For example if a strip of metal zinc (Zn$_{(s)}$) is placed into a solution containing copper ions (Cu^{+2} $_{(aq)}$), the zinc atoms on the surface of the metal strip will give up their electrons to the copper ions in solution. The oxidized zinc ions (Zn^{+2}$_{(aq)}$) will dissolve, and the reduced copper atoms (Cu$_{(s)}$) will precipitate as copper metal (Figure 6-15).

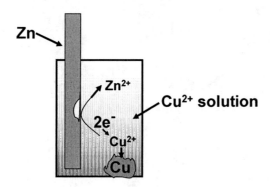

Figure 6-15
Metallic zinc gives two electrons to Cu^{2+}, the zinc is oxidized to Zn^{+2} and dissolves while the Cu^{+2} is reduced to Cu solid and precipitates

If we separate the copper ions from the zinc metal by placing them in different containers, but connecting them with a conductive wire, the electrons will flow through the wire to get from the zinc to the copper ions producing an electric current (Figure 6-16).

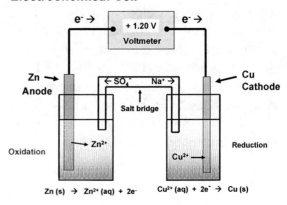

Figure 6-16
Voltaic or electrochemical cell

The **electrochemical cell** pictured in Figure 6-16 is a device in which a redox reaction is used to produce an electric current. The chemical equation representing the redox reaction in the cell above is written as:

$$Zn(s) + Cu^{2+}(aq) \rightarrow Zn^{2+}(aq) + Cu(s)$$

Note that Zn(s) is oxidized and Cu2+ is reduced

Most electrochemical cells are made using two metal electrodes immersed in different ionic solutions containing the ions of the individual metals. **Electrodes** are materials through which electrons enter and leave an electrochemical cell. In the cell pictured above the zinc electrode is placed in a solution of zinc ions and the copper electrode is placed in a solution of copper ions. The electrode, where oxidation occurs, is called the **anode**, and the **cathode** is the electrode where reduction occurs. The two electrodes are connected by a wire, allowing the electrons generated at the anode to travel to the cathode and produce the electric current. This current of electrons is measured by adding a volt meter between the two electrodes. In Figure 6-16, the electrochemical cell is producing an electric current of +1.20 volts. Finally, a salt bridge is used to keep both solutions from acquiring a net positive or negative charge. As a result of the electrons being produced at the anode, an increase of Zn^{+2} ions in the zinc solution occurs. This increase in positive charge is counter balanced by negatively charged ions entering the solution from the salt bridge. As electrons enter the cathode, the copper ions are reduced to copper atoms and, as a result, the solution gains a net negative charge. Positive ions from the salt bridge enter the solution, canceling this effect and keeping the solution electrically neutral. Keeping the solutions electrically neutral allows the electrons to continuously flow from the anode to cathode. Without the salt bridge, the flow of electrons would stop. As the zinc solution containing the anode becomes more positive, since electrons are attracted to a positive charge, they will enter this solution rather than cross to the cathode. The usable life of this type of an electrochemical cell is limited. As the zinc strip slowly dissolves and the copper ions are deposited as copper atoms on the cathode, the ability of the cell to produce an electric current diminishes. This is the case with most disposable batteries. If, however, the reaction were easily reversed by dissolving the copper atoms and depositing the zinc ions as zinc atoms, the battery could be reused. Batteries that are not reusable are called primary, and those in which the redox reaction can be reversed are called secondary batteries.

Primary Batteries

The most widely used types of primary cells are the dry cell batteries. These batteries differ in various ways, but all are types of electrochemical cells. Therefore, every dry cell has an anode and a cathode in contact with an ionic solution, promoting the flow of electrons. The major types of dry primary batteries are carbon-zinc cells and alkaline cells.

Carbon-Zinc Cells

The all-purpose batteries used in flashlights, toys, cameras, etc. are carbon-zinc cells. These cells are contained within a zinc "can," which serves both as a container for the parts of the cell and as the anode (Figure 6-17).

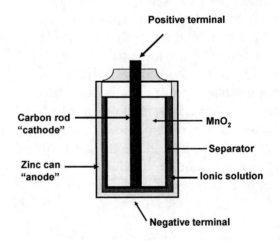

Figure 6-17
Carbon-zinc primary battery

A carbon rod in the center of the cell functions as the cathode. The actual cathode material, however, is a mixture of manganese dioxide and carbon powder packed around the rod. The ionic solution is a paste composed of ammonium chloride, zinc chloride, and water. The electrodes are separated by a sheet of porous material, such as paper or cardboard, and soaked with the ionic solution that prevents the electrode materials from mixing together and reacting when the battery is not being used. Without a separator, the zinc anode could wear away prematurely and reduce the life of the battery. The electrochemical reaction producing the current is very similar to the one shown in Figure 6-16. The zinc atoms at the surface of the anode are oxidized, becoming zinc ions leaving their electrons on the anode. The anode thus gains an excess of electrons and becomes more negatively charged than the cathode. If the cell is connected to an external circuit, the zinc anode's excess electrons flow through the circuit, producing an electric current, to the carbon rod cathode. As the electrons enter the cell through the rod, they combine with molecules of manganese dioxide and molecules of water. As these substances are reduced and react with one another, they produce manganese oxide and negative hydroxide ions. This reaction makes up the second half of the cell's discharge process. It is accompanied by a secondary reaction in which the negative hydroxide ions combine with positive ammonium ions that form ammonium chloride, producing molecules of ammonia and molecules of water.

The chemical reactions that produce electricity inside a carbon-zinc cell continue until the manganese dioxide wears away. Once this cathode material has been "used up," the cell can no longer provide useful energy and is "dead."

Alkaline cells resemble carbon-zinc cells and undergo similar chemical reactions. But the two types differ in several important ways. An alkaline cell has a highly porous zinc anode that oxidizes more readily than that of a carbon-zinc cell. Its ionic solution is a strong alkaline solution called potassium hydroxide. This compound conducts electricity inside the cell better than does the solution of ammonium chloride and zinc chloride in a carbon-zinc cell, enabling an alkaline cell to deliver higher currents than a carbon-zinc cell.

Alkaline cells serve as an excellent power source for electric shavers, portable TVs, and radios. They are more economical than zinc-carbon cells because they last from five to eight times as long.

Secondary Batteries

Secondary batteries are made so that their chemical reactions can be reversed. This feature enables them to be recharged if their electric current has stopped. The most common types of secondary batteries are lead-acid storage batteries and nickel-cadmium batteries.

Lead-acid storage batteries consist of a plastic or hard-rubber container that holds three or six electrochemical cells, each of which has two sets of lattice like electrodes called plates. The frames of these structures, called grids, are made of a lead-antimony alloy. The spaces between the anodes are filled with a mass of pure lead. The spaces between the cathodes contain lead dioxide. An ionic solution of sulfuric acid and water surrounds the electrodes (Figure 6-18).

When an electric current is produced, the chemical reactions takes place between the electrode materials and the ionic solution. At the anode, atoms of pure lead react with negative sulfate ions of the solution. As the lead atoms combine with the sulfate ions, each lead atom loses two electrons and becomes a molecule of lead sulfate. The electrons lost by the lead atoms flow from the anode to the cathode, producing an electric current. At the cathode, they combine with molecules of lead dioxide which, in turn, combine with the hydrogen and sulfate ions of the ionic solution, producing lead sulfate and water.

Lead Storage Battery

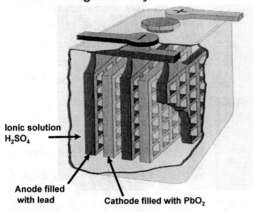

Figure 6-17
Lead Storage Battery

The current-producing process decreases and dilutes the electrolyte of sulfuric acid by using up sulfate ions and by adding water molecules to the solution. The battery becomes discharged when so little sulfuric acid remains that the necessary chemical reactions can no longer occur.

A lead-acid battery can be recharged by means of a battery charger, which forces electrons through the battery in a direction opposite to that of the discharge process. This action reverses the chemical reactions that occur at the electrodes when a battery discharges. The reversed reactions of the charging process restore the electrode materials to their original form. They also increase the amount of sulfuric acid in the electrolyte to a satisfactory level. Once recharged, a lead-acid battery can again produce electricity. Lead-acid storage batteries produce energy for the electrical systems of almost all automobiles.

Nickel-cadmium storage batteries operate on the same general principles as the lead-acid batteries but use different chemical substances. In a nickel-cadmium battery, the negative electrode is made of cadmium and the positive electrode of nickel oxide. A solution of potassium hydroxide serves as the electrolyte.

The chemical composition of a nickel-cadmium battery allows the battery container to be sealed airtight, which prevents the corrosive electrolyte from leaking. Because of this advantage, nickel-cadmium batteries are used in most portable tools like cellular phones, etc. Most space satellites also use these batteries.

Corrosion

Millions of dollars are lost each year because of the oxidation or corrosion of metals. Although many other metals corrode, most of this loss is due to the corrosion of iron and steel. The problem with iron is that the oxide it forms does not adhere to the surface of the metal and flakes off, causing "pits" to form. Extensive pit formation eventually causes structural weakness and disintegration of the metal. This, however, does not happen to metals like aluminum and copper which upon oxidation, form a very stable oxide coating which strongly bonds to the surface of the metal and prevents these surfaces from further corrosion.

Corrosion occurs in the presence of moisture. For example, when iron is exposed to moist air, it reacts with oxygen to form iron (III)[6] oxide, or rust.

$Fe_2O_3 \bullet XH_2O$

The amount of water complexed with the iron (III) oxide (ferric oxide) varies as indicated by the letter "X". The amount of water present also determines the color of rust, which may vary from black to yellow to orange brown. The formation of rust is a very complex process which is thought to begin with the oxidation of iron to ferrous (Fe^{+2}) ions.

$$Fe \rightarrow Fe^{+2} + 2\,e^-$$

Both water and oxygen are required for the next sequence of reactions. The ferrous (Fe^{+2}) ions are further oxidized to form ferric ions (Fe^{+3}) ions.

$$Fe^{+2} \rightarrow Fe^{+3} + 1\,e^-$$

The electrons provided from both oxidation steps are used to reduce oxygen.

$$O_{2\,(g)} + 2\,H_2O_{(l)} + 4\,e^- \rightarrow 4\,OH^-$$

The ferric ions (Fe^{+3}) then combine with oxygen to form ferric oxide (iron (III) oxide) which is then hydrated with varying amounts of water. The overall equation for the rust formation may be written as:

$$4\,Fe^{2+}_{(aq)} + O_{2\,(g)} + X\,H_2O_{(l)} \longrightarrow 2\,Fe_2O_3 \cdot XH_2O_{(s)} + 8\,H^+_{(aq)}$$

The formation of rust can occur at some distance away from the actual pitting or erosion of iron. The electrons produced from the initial oxidation of iron can be conducted through the metal and Fe^{+3} ions can diffuse through the water layer to another point on the metal surface where oxygen is available. This process is the same as found in an electrochemical cell in which iron serves as the anode, oxygen gas as the cathode, and the aqueous solution of ions as a "salt bridge"(Figure 6-18).

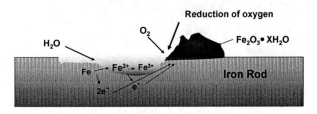

Figure 6-18
Rusting of an iron surface

---

[6] The Roman numeral three indicates that iron has a positive three (+3) charge.

Chapter 6 Exercises

1. For the equation, $2 NaN_3(s) \rightarrow 2 Na(s) + 3 N_2(g)$, which statement is false?

    a. the equation is balanced

    b. the products are sodium and nitrogen

    c. the sum of all coefficients is 7

    d. the products are all in the same state of matter

2. For the equation,

$$H_2(g) + Cl_2(g) \rightarrow 2 HCl(g),$$

    which statement is false?

    a. 1 mol of $Cl_2(g)$ is a reactant

    b. 2 molecules of HCl are formed

    c. 4 grams of $H_2(g)$ reacts with 1 mol of $Cl_2(g)$

    d. the sum of the coefficients of the reactants is 2

3. Which is necessary for a chemical reaction to occur?

    a. reactant molecules must collide

    b. collisions must be of a certain energy

    c. neither a nor b

    d. both a and b

111

4. What factors determine the rate of a reaction?

    a. nature of reactants

    b. concentration of reactants

    c. activation energy

    d. all of these

5. Define Catalyst

    _____

    _____

    _____

    _____

6. Preserving food by freezing controls reaction rates by

    a. decreasing the number of collisions

    b. increasing concentration

    c. changing the number of effective reactant collisions

    d. both a and c are correct

7. Define the First Law of Thermodynamics?

    _____

    _____

    _____

    _____

8. Which will increase the rate of a reaction?

    a. increasing temperature

    b. increasing temperature and concentration

    c. increasing concentration of reactants

    d. all of these

9. If you add more $CO_2(g)$ to an equilibrium mixture of $CaCO_3(s)$, $CaO(s)$ and $CO_2(g)$ in a hot closed vessel, which way would you expect the equilibrium to shift ?

$$CaCO_3(s) \leftrightarrow CaO(s) + CO_2(g)$$

a. to form more reactant

b. to form more product

c. no change

d. indeterminable based in information given

11. Calculate the molar mass of H_2SO_4?

_____g/mol

12. How many grams are in 3.0 moles of H_2SO_4?

_____grams

13. How many moles are in 90 grams of H_2O?

_____moles

14. Which will have the largest molar mass?

a. 1 mol of Ar, argon

b. 1 mol of CO, carbon monoxide

c. 1 mol of N_2, nitrogen

d. 1 mol of Sn, tin

113

15. Define the second law of thermodynamics,

_____

_____

_____

_____

16. Define entropy:

_____

_____

_____

_____

17. According to the equation below if 4 moles of hydrogen react with an unlimited supply of chlorine how many moles of HCl will form?

$$H_2(g) + Cl_2(g) \rightarrow 2\ HCl(g)$$

_____moles of HCl

18. What is the difference between exothermic and endothermic reactions?

_____

_____

_____

_____

19. Consider the equation below.

$$Zn(s) + 2\ HCl(aq) \rightarrow ZnCl_2(s) + H_2(g)$$

What weight of $ZnCl_2$ will be produced from 1.00 mole of Zn reacting with excess HCl?

_____grams of $ZnCl_2$

114

20. A chemical reaction that releases heat during the reaction is

a. exothermic

b. endothermic

c. a source of quick heat

d. impossible to reverse

21. Balance the following equation?

$$CO_2 + H_2O \rightarrow C_6H_{12}O_6 + O_2$$

22. Le Chatelier's principle says

a. equilibrium mixtures are changed only when reactants are added

b. equilibrium mixtures will change to form more products when a reactant is added

c. the balance between reactants and products is not altered by changing the product concentration

d. none of these

23. How will increasing the temperature affect the rate of a chemical reaction?

    a. never increase

    b. always decrease

    c. usually increase

    d. usually decrease

24. How does changing concentration affect the rate of a chemical reaction?

    a. increasing the concentration of a reactant will increase the reaction rate

    b. the concentration has very little effect on reaction rates

    c. the reaction rate increases only when all reactant concentrations are increased

    d. increasing the concentration of products increases reaction rates

25. Chemical equilibrium exists for a chemical reaction when

    a. equal amounts of chemicals exist on reactant and product sides

    b. one or more of the reactants are used completely

    c. when the forward and reverse reaction rates are equal

    d. none of these

26. What is the weight of one mole of water molecules, H_2O?

    _____g H_2O

27. How many moles of phosphorous are in one mole of $Ca_3(PO_4)_2$?

    _____ moles of P

28. Which of the following samples has the most entropy?

a. 1 mol solid hydrogen

b. 1 mol liquid hydrogen

c. 1 mol hydrogen gas

d. all are the same because they are all hydrogen

29. How many atoms of nitrogen are in one mole of nitrogen dioxide, NO_2?

_____atoms of N

30. Define reduction and oxidation.

_____

_____

_____

_____

31. Assign oxidation number to all the elements in the following:

$$CH_4 \text{ (g)} + 2O_2 \text{ (g)} \rightarrow CO_2 \text{ (g)} + 2H_2O \text{ (}l\text{)}$$

32. Assign oxidation number to all the elements in the following:

$$SnO_2\text{(g)} + 2\text{ C(s)} \rightarrow Sn\text{(s)} + 2\text{ CO(g)}$$

33. In the equation

$$2\,Na(s) + Cl_2(g) \rightarrow 2\,Na^+ + 2\,Cl^-,$$

what species is oxidized? _____

34. In the equation

$$SnO_2(g) + 2\,C(s) \rightarrow Sn(s) + 2\,CO(g),$$

what substance is the reducing agent? _____

35. In the equation

$$2\,Na(s) + Cl_2(g) \rightarrow 2\,Na^+ + 2\,Cl^-$$

what substance is the oxidizing agent? _____

36. In the equation

$$SnO_2(g) + 2\,C(s) \rightarrow Sn(s) + 2\,CO(g),$$

what substance is the reduced? _____

37. Define free radicals:

_____

_____

_____

_____

118

38. Define reducing agents:

_____

_____

_____

_____

39. Antioxidant vitamins

    a. react with DNA

    b. donate electrons to free radicals

    c. are reduced

    d. form free radicals in the body

40. Describe how an electrochemical cell works.

_____

_____

_____

_____

_____

_____

41. Which is not an example of a primary battery?

    a. lead-acid automobile battery

    b. alkaline battery

    c. carbon-zinc cells

    d. alkaline cells

42. $PbSO_4$, Pb, and H_2SO_4 are chemicals in

    a. dry cell

b. primary battery

c. salt bridge

d. automobile battery

43. Which of the following is not a definition of oxidation?

a. loss of electrons

b. loss of hydrogen

c. loss of protons

d. addition of oxygen

44. Which of the following is not a definition of reduction?

a. gain of electrons

b. gain of protons

c. gain of hydrogen

d. loss of oxygen

45. Which equation does not fit the definition of oxidation?

a. $Ag^+ + e^- \rightarrow Ag$

b. $C_2H_6 \rightarrow H_2 + C_2H_4$

c. $C + O_2 \rightarrow CO_2$

d. $Na \rightarrow Na^+ + e^-$

46. Which of the following is a reduction of carbon?

a. $C + 2\,H_2 \rightarrow CH_4$

b. $C + O_2 \rightarrow CO_2$

c. $2\,C + O_2 \rightarrow 2\,CO$

d. $C^{2+} \rightarrow C^{4+} + 2e^-$

47. Which of the following is not necessary for corrosion of iron?

a. iron   b. water   c. light   d. oxygen

48. Which of the following might prevent corrosion of iron?

a. a water spray

b. a coating of a rust inhibitor

c. acid treatments

d. salt solutions

49. Oxidation occurs at the _____ in an electrochemical cell.

50. A rechargeable battery is which pf the following:

a. primary battery

b. dry cell

c. electroplating cell

d. secondary battery

51. Which of the batteries listed below is a reusable battery?

a. dry cell battery

b. carbon-zinc cells

c. alkaline battery

d. lead storage battery

Chapter 7

Reactions of Acids and Bases

Acids and bases are a class of compounds that have been studied for centuries. The reactions of acids and bases with other materials and with each other are numerous and are very common in our natural surroundings. Acid-base chemistry, as it is sometimes called, plays a major role in the vitality of all living creatures. The level of acid in the blood, for instance, is kept constant by the reactions of acids and bases. If the acid level increases or decreases by only a fraction serious health problems will arise. Acids and bases are found abundantly in nature. They are found in almost all fruits and vegetables and are characterized by their unique tastes. Acids and bases are also very important to industry. Sulfuric acid, for example, is used so extensively that more than 60 million tons of it was produced in the United States in 2003. It has widely varied uses and plays some part in the production of nearly all manufactured goods. Also, there are many compounds containing acids and bases that we use every day in our homes.

In this chapter we will consider the properties of acids and bases, the reactions they undergo and some important uses for them. Some key questions we will address are:

What are the properties of acids and bases?

What is a neutralization reaction?
What is pH?

What are buffers?

What are some uses for acids and bases?

Figure 7-1
The sour taste of lemons is due to the presence of an acid

Acid-Base Definitions

Despite the familiarity and the importance of acids and bases, it has not been a simple matter for scientists to exactly define them. An early definition of acids and bases was based solely on their observable properties. For example, solutions of acids, like vinegar and lemon juice, taste sour. They were observed to change the color of certain vegetable dyes. Acids also dissolve some metals releasing hydrogen gas. Bases on the other hand form solutions having a bitter taste and feel slippery to the touch. Like acids, bases also cause certain vegetable dyes to change color. And finally, salts were observed to form when acids and bases were allowed to react. Today the Arrhenius and Bronsted-Lowery definitions of acids and bases are the two most widely used.

Arrhenius Acid-Base

123

Although useful, these early definitions of acids and bases do not relate their observable properties to their compositions and molecular structures. Svante Arrhenius suggested the first structural acid-base definition; where an **Arrhenius acid** is any substance that produces a hydrogen ion (H^+) when dissolved in water, the **Arrhenius base** is any substance that produces the hydroxide ion (OH^-) when dissolved in water. For example, when hydrogen chloride is dissolved in water, it produces the hydrogen ion:

$$HCl(aq) \rightarrow H^+(aq) \ + \ Cl^-(aq)$$

And when sodium chloride dissolves in water, it produces the hydroxide ion:

$$NaOH(aq) \rightarrow Na^+(aq) \ + OH^-(aq)$$

The only problem with this definition is that the hydrogen ion produced from an acid is not in the form of a bare proton as Arrhenius suggests. Instead the proton is attached to a water molecule to form a **hydronium ion,** (H_3O^+). Therefore a more accurate equation for the dissolving of HCl in water would be:

$$HCl(aq) + H_2O(l) \rightarrow H_3O^+(aq) + Cl^-(aq)$$

Bronsted-Lowry Acid-Base

The Bronsted-Lowry acid-base definition is more encompassing than that by Arrhenius. Any substance that is an acid or a base, by the Arrhenius definition, will still be an acid or a base according to Bronsted-Lowry. However, the Bronsted-Lowry definition will include some substances as acids and bases, which would not be included by the Arrhenius definition.

According to **Bronsted-Lowry,** an **acid** is a proton (H^+) donor and a **base** is a proton acceptor. The Bronsted-Lowry definition requires that in order for a substance to act as an acid, there must be a base present, because if a substance is to donate a proton, there must be another substance present to accept it. So an acid-base reaction is simply the transfer of a proton from an acid to a base. For example, the reaction of HCl and ammonia:

$$HCl_{(aq)} + NH_{3\,(aq)} \rightarrow \ NH_4^+{}_{(aq)} + H_3O^+{}_{(aq)}$$

Table 7-1 lists the properties of acids and bases, taking into account all of the current definitions.

Acids	Bases
Sour taste	Bitter taste
Give H^+ ions (Arrhenius)	Provide OH^- (Arrhenius)
Donate a proton (Bronsted-Lowry)	Accept a proton (Bronsted-Lowry)
React with metals to give Hydrogen	Slippery feeling
Examples; vinegar, tomatoes, citrus fruit, and aspirin	Examples; ammonia, baking soda, soap, and detergents

Table 7-1
Properties of acids and bases

Strength of Acids and Bases

The strengths of acids and bases are based solely on the extent to which they produce the hydronium and hydroxide ions, respectively, when dissolved in water. By definition a **strong acid** com-

124

pletely dissociates in water by donating 100% of its available hydrogen ions producing hydronium ions (H₃O⁺). For example, when hydrogen chloride is dissolved in water it will completely break apart and produce H₃O⁺ and Cl⁻;

HCl (aq) + H₂O (l) $\xrightarrow{100\%}$ H₃O⁺ (aq) + Cl⁻ (aq)
0% in solution   100% in solution

Likewise, a **strong base** completely dissociates in water, producing 100% of the potential hydroxide ions (OH⁻). Take sodium hydroxide for example:

NaOH (aq) $\xrightarrow{100\%, H_2O}$ Na⁺ (aq) + OH⁻ (aq)
0% in solution   100% in solution

On the other hand, weak acids and bases do not dissociate completely in water, but exist in a state of equilibrium as both the bonded and unbound forms. Hydrogen fluoride is a weak acid and, therefore, exists in solution as the HF molecule, the fluoride ion (F⁻), and the hydronium ion (H₃O⁺). The extent of dissociation is on the order of only about 20%, meaning that 80% of hydrogen fluoride exists as molecular HF in solution leaving the production of the hydronium ion to only 20%.

HF (aq) + H₂O (l) ↔ F⁻ (aq) + H₃O⁺ (aq)
80% In solution   20% in solution

Ammonia is a weak base and exists in water as free ammonia (NH₃), the ammonium ion (NH₄⁺); as a result it produces a small amount of the (OH⁻) ion.

NH₃ (aq) + H₂O (l) ↔ NH₄⁺ (aq) + OH⁻ (aq)
85% In solution   15% in solution

In this case, only 15% of ammonia dissociates to form the ammonium ion (NH₄⁺).

Table 7-2 lists the common strong acids and bases.

Acid Name	Formula	Base Name	Formula
Hydrochloric acid	HCl	Sodium Hydroxide	NaOH
Nitric acid	HNO₃	Potassium Hydroxide	KOH
Hydrobromic acid	HBr	Lithium hydroxide	LiOH
Sulfuric acid	H₂SO4	Magnesium hydroxide	Mg(OH)₂
Hydroiodic acid	HI	Calcium hydroxide	Ca(OH)₂

Table 7-2
Table of strong acids and bases

Neutralization Reactions of Strong Acids and Bases

One of the fundamental chemical properties of strong acids and strong bases is that they neutralize or destroy one another. A **neutralization reaction** results when equal amounts of a strong acid and base react, forming an ionic compound and water. The ionic compound that forms is called a **salt.** Most ionic compounds, with the exception of hydroxides and oxides, are salts.

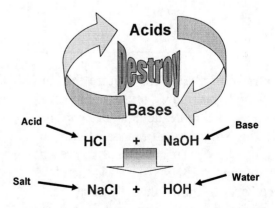

For example, when equal amounts of hydrochloric acid (HCl (aq)) and sodium hydroxide (NaOH (aq)) react, the salt sodium chloride and water are formed.

HCl (aq) + NaOH (aq) → NaCl (aq) + H$_2$O (aq)

Example 7-1

Determine the products formed from the reaction of nitric acid and lithium hydroxide

Answer

Nitric acid HNO$_3$ will give up its hydrogen ion (H$^+$) completely.

HNO$_3$ → H$^+$ + NO$_3^-$

Likewise sodium hydroxide NaOH will completely lose its hydroxide ion (OH$^-$)

NaOH → Na$^+$ + OH$^-$

This results in the hydrogen ion and the hydroxide ion combining to form water, and the sodium ion, and the nitrate ion forming sodium nitrate

HNO$_3$ → H$^+$ + NO$_3^-$

NaOH → Na$^+$ + OH$^-$

   NaNO$_3$ H$_2$O
 Sodium nitrate water

Neutralization reactions require that the acid and base are present in equal amounts. If more acid is present, the resulting solution will not be neutral but acidic; likewise, if more base is present the solution will be basic. It is therefore necessary to know when equal amounts of acid and base are present in a solution. **Indicators** are substances that change color depending on the acidity or basicity of the solution. They are one color in a basic solution and another in an acidic solution. For example, the indicator phenolphthalein is clear in an acidic solution but changes to pink in a basic solution (Figure 7-2).

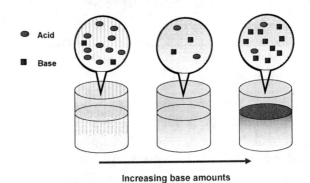

Figure 7-2
    Phenolphthalein changes from clear to pink when the solution changes from acidic to basic

When a small amount of the indicator phenolphthalein is mixed with a solution of a strong acid, and a strong base is slowly added, neutralization occurs the instant a color change is observed. If, however, the base is added too rapidly, the risk of passing the neutralization point and producing a basic solution increases. If this happens a strong acid can be slowly added until the pink color just disappears and the solution becomes clear.

Molarity- Concentrations of Acids and Bases

The amount of dissolved substance in a given amount of water is called its *concentration*. Scientists frequently use the unit of molarity to define the concentration of any substance in a solution. **Molarity (M)** is defined as the number of moles of a substance in one liter of solution. A **solution** is defined as a mixture of a **solute**, the dissolved substance, and the solvent, what the solute is dissolved in. For example, if 1.50 moles of HCl were dissolved in enough water to make exactly 1.0 liters of solution, the resulting concentration would be 1.50 molar[7] (more commonly written as 1.50 M).

$$\frac{1.50\ \text{moles of solute}}{1.0\ \text{liters of solution}} = 1.50\ \text{M}$$

Example 7-2

What is the molarity of a solution prepared by dissolving 1.25 moles of sodium hydroxide in enough water to make 350 ml of solution?

Answer

The definition of molarity is the number of moles of solute divided by the liters of solution.

$$\frac{\#\ \text{of moles of solute}}{\#\ \text{of liters of solution}} = \text{Molarity}$$

There are 1.25 moles of solute (sodium hydroxide)

The volume of the solution in liters is

$$350ml\left(\frac{1\ \text{liter}}{1000\ \text{ml}}\right) = 0.350\ \text{liters}$$

Therefore:

$$M = \frac{1.25\ \text{moles of NaOH}}{0.350\ \text{liters of solution}} = 3.57\ \text{M}$$

The solution is 3.57 molar NaOH

Example 7-3

---

[7] Molar is short for molarity

Calculate the molarity of a solution made by dissolving 3.5 g (mw=40 g/mol) of NaOH in enough water to make 1.6 liters of solution.

Answer

Here, the number of grams of NaOH is given instead of the number of moles. And since molarity is the number of moles per liter, we have to first convert the number of grams of NaOH to moles of NaOH.

$$3.5\ \text{grams NaOH}\left(\frac{1.0\ \text{mol NaOH}}{40\ \text{grams NaOH}}\right)$$

$$= 0.0875\ \text{NaOH}$$

Now we can use the definition of molarity:

$$M = \frac{0.0875\ \text{mol of NaOH}}{1.6\ \text{liters of solution}} = 0.055\ \text{M}$$

This solution is 0.055 molar NaOH

Example 7-3

Calculate the concentration of the hydroxide ion in the solution found in Example 7-3.

Answer
By definition, NaOH is a strong base and therefore, dissociates completely:

$$\text{NaOH} \xrightarrow{100\%} \text{Na}^+ + \text{OH}^-$$

This means that the OH⁻ concentration is the same as the NaOH concentration; 0.055 M = concentration of [OH⁻]

Brackets are used to represent the concentration of a particular substance. For example to represent the concentration of the hydroxide ion as 0.055 M, we would write $[OH^-]$ = 0.055 M; this is read as, the concentration of the hydroxide ion is 0.055 molar.

Example 7-4

Calculate the hydronium ion concentration of 1.25 liters of a solution containing 12.5 grams of nitric acid (HNO_3 mw= 63.0 g/mol)

Answer

Since nitric acid is a strong acid and therefore completely dissociates in water;

$$HNO_3 + H_2O \xrightarrow{\ 100\%\ } H_3O^+ + NO_3^-$$

The $[H_3O^+]$ is equal to the $[HNO_3]$

The $[HNO_3]$ is therefore determined from the number of moles of HNO_3 divided by the volume of the solution in liters

$$12.5\,\text{g HNO}_3 \left(\frac{1\,\text{mole HNO}_3}{63\,\text{g HNO}_3} \right)$$
$$= 0.198\,\text{mol HNO}_3$$

And:

$$[HNO_3] = \frac{0.198\,\text{mol HNO}_3}{1.25\,\text{liters of solution}}$$

$$= 0.158\,\text{M}$$

Therefore:

$$[HNO_3] = 0.158\,\text{M} = [H_3O^+]$$

The pH Scale

The most common measure of the acidity and basicity of a solution is the pH. Quite simply, pH is defined as the negative logarithm of the hydronium ion concentration and is written as;

$$pH = -\log [H_3O^+]$$

For example, the pH of a strong acid solution having a hydronium ion concentration of 0.035 M is therefore;

$$pH = -\log(0.035) = 1.46$$

As stated earlier, the difference in a basic and an acidic solution is the relative amounts of acid to base in the mixture. When a ratio of acid to base equals one the solution is neutral--neither acidic nor basic. Recall that, when referring to strong acids and bases, the concentration of the acid is equal to the concentration of the hydronium ion, $[H_3O^+]$, and the concentration of a strong base is the same as the concentration of the hydroxide ion $[OH^-]$. Therefore, within a neutral solution the hydronium ion concentration is equal to the hydroxide concentration.

Neutral solution = $[H_3O^+]$ = $[OH^-]$

If the concentration of the hydronium ion in pure water (a neutral solution) were determined, the concentration of the hydroxide ion would also be known, and it would be equal to that of the hydronium ion. The concentration of $[H_3O^+]$ in pure water has been accurately measured and given as 1.0×10^{-7} M. Therefore in a neutral solution the concentration of both $[H_3O^+]$ and $[OH^-]$ is 1.0×10^{-7} M.

Using the hydronium concentration in pure water, its pH is therefore 7.00.

$$pH_{\text{pure water}} = -\log(1.0 \times 10^{-7}) = 7.00$$

An acidic solution will therefore have a $[H_3O^+]$ greater than 1.0×10^{-7} M, and a basic solution will have a $[H_3O^+]$ less than 1.0×10^{-7} M. Also, as the concentration of the hydronium ion in a solution increases, the pH of the solution decreases. Conversely, as the hydronium ion in a solution decreases, the pH of the solution increases. For instance, if a solution has a $[H_3O^+]$ of 1.3×10^{-4}, its pH is 3.89 (acidic), and if the concentration is 1.3×10^{-9}, its pH is 8.89 (basic).

The standard pH scale has a range from 0.00 to 14.00, where acidic solutions have a pH between 0.00 to just below 7.00 and basic solutions from just above 7.00 to 14.00. The pH of several common materials are given in Table 7-3

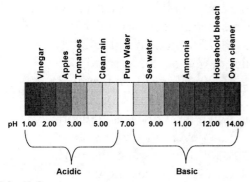

Table 7-3
pH of various materials

The relationship between pH and $[H_3O^+]$ is therefore inversely and logarithmic. Put another way, as the pH decreases by one unit, the hydronium ion concentration increases by ten times. Conversely, as the pH increases by one unit the hydronium ion concentration decreases ten times. Table 7-3 summarizes the relationship between pH and $[H_3O^+]$.

$[H_3O^+]$	pH	Decrease in $[H_3O^+]$ or Increase in $[OH^-]$
1.0×10^{0}	0	
1.0×10^{-1}	1	10 x
1.0×10^{-2}	2	100 x
1.0×10^{-3}	3	1000 x
1.0×10^{-4}	4	10000 x
1.0×10^{-5}	5	100000 x
1.0×10^{-6}	6	1000000 x
1.0×10^{-7}	7	10000000 x
1.0×10^{-8}	8	100000000 x
1.0×10^{-9}	9	1000000000 x
1.0×10^{-10}	10	10000000000 x
1.0×10^{-11}	11	100000000000 x
1.0×10^{-12}	12	1000000000000 x
1.0×10^{-13}	13	10000000000000 x
1.0×10^{-14}	14	100000000000000 x

Table 7-3
Relationship between pH and the concentration of $[H_3O^+]$; the factor at which the $[H_3O^+]$ concentration decreases is the same factor at which the $[OH^-]$ concentration increases

---

Example 7-5

When a 12 oz can of Pepsi© is poured into a glass containing the same volume of water the pH of the water changes from 7.00 to 4.00. By what factor does the hydronium ion concentration increase in the water?

Answer

The pH has decreased by 3 pH units from 7.00 to 4.00. Since each unit decrease in pH results in a 10 fold increase in $[H_3O^+]$ concentration. We can therefore say the increase in hydronium concentration is;

$10 \times 10 \times 10 = 10^3 = 1000$ times

Since, in the previous example, the concentration of the hydronium ion increased by a factor of a thousand, the hydroxide ion concentration must have decreased by the same factor.

Buffers and pH

It is quite easy to change the pH of some solutions by either adding an acid or a base (Example 7-5). The pH of other solutions however, is resistant to change when an acid or base is added; solutions that show this kind of behavior are called **buffers**. The pH-resistant characteristics of buffers are due to their abilities to react with and remove the added acid or base from a solution. A buffers ability to react with both acids and bases is attributed to its composition. Buffers are solutions of either weak acids or weak bases and their conjugate partners. A **conjugate partner** is the product formed, other than $[H_3O^+]$ or $[OH^-]$, from the dissociation of either a weak acid or weak base. For instance when the weak acid HF is dissolved in water the products are the fluoride ion and the hydronium ion;

$$HF + H_2O \rightarrow F^- + H_3O^+$$

HF and F- are conjugate partners. Therefore, a solution of HF and F⁻ results in a buffered solution.

Buffers are very important types of solutions used extensively in both the industrial and medical felids. The roles they play in our bodies, however, are by far more important. Blood, for instance, is a buffered solution composed of many things including red and white blood cells, plasma, and other important biological components. The buffering components found in blood are carbonic acid (H_2CO_3) and the bicarbonate ion (HCO_3^-). In your blood, these weak ac-

id-base partners will react with any added acid or base as follows:

When an acid is added:

$$H_3O^+ + HCO_3^- \rightarrow H_2CO_3 + H_2O$$

When a base is added:

$$OH^- + H_2CO_3 \rightarrow HCO_3^- + H_2O$$

The bicarbonate ion reacts with any added acid, producing water and carbonic acid, whereas the carbonic acid reacts with any added base, producing water and the bicarbonate ion. In both cases, the acid and base are completely removed. This action keeps the pH of blood at 7.40.

The ability of blood to maintain a pH of 7.40 is extremely important, because very small changes will result in life threatening conditions. For instance, **acidosis** occurs when the pH of blood drops below 7.35, making it slightly more acidic; during acidosis, a drop below a pH of 6.80 will result in death. A person affected with acidosis experiences symptoms like drowsiness and disorientation and can even go into a state of unconsciousness or coma. On the other hand, if blood pH becomes more basic, or alkaline, the condition is called alkalosis. **Alkalosis** occurs when the blood pH rises above 7.45, but a rise above 8.0 also results in death. Symptoms of alkalosis include light-headedness, agitation, and dizziness.

Acid Rain

Any precipitation that has a measured pH of 5.60 or less is called **acid rain**. A more precise name would be *acid precipitation* since it can be in the form of snow, fog, mist or even dry particles. Acid rain is formed when sulfur dioxide (SO$_2$), and nitrogen monoxide (NO) combine with oxygen and moisture in the atmosphere and make sulfuric, nitric and nitrous acids.

Sulfur dioxide is a colorless gas formed as a by-product from the combustion of fossil fuels used in industrial processes, such as the production of iron, steel, utility factories, and crude oil processing. These Industrial processes are responsible for about 76 percent of the sulfur dioxide emitted into the air. Combustion of fossil fuels used for commercial and residential purposes account for about 18% and those combusted for transportation purposes emit about 5.0 percent, while natural occurring events like volcanoes and forest fires contribute roughly 1 percent (Figure 7-3).

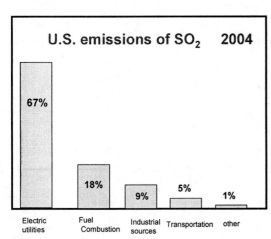

Figure 7-3
SO$_2$ emissions in the U.S. in 2004

Sulfur dioxide produces acid rain when it reacts with moisture found in the atmosphere. When this happens, sulfur dioxide immediately oxidizes to form a sulfur trioxide (SO$_3$).

SO$_2$ (g) + O$_2$ (g) → SO$_3$ (g)

The sulfur trioxide then reacts with water forming sulfuric acid.

SO$_3$ (g) + H$_2$O (*l*) → H$_2$SO$_4$ (aq)

Acid rain is then produced when moisture containing sulfuric acid precipitates in the form of snow, rain, fog or mist.

Nitrogen monoxide, the other contributor of acid rain, is primarily produced from the combustion of fossil fuels. Automobiles are responsible for about 54 percent of production of this gas. While 44 percent comes from the combustion of fossil fuels from industry, electric utilities, commercial and residential use. And only 2 percent is generated by natural processes such as forest fires, volcanic action, and lightning (Figure 7-4).

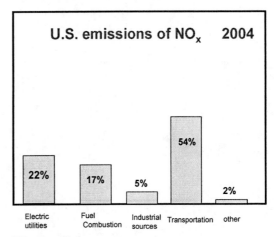

Figure 7-4
NO and NO$_2$ emissions in the U.S. in 2004

Like sulfur dioxide, nitrogen monoxide rises into the atmosphere reacting slowly with oxygen producing nitrogen dioxide (NO$_2$).

2NO (g) + O$_2$ (g) → 2 NO$_2$ (g)

Nitrogen dioxide then reacts with water to form nitrous acid (HNO$_2$) and nitric acid (HNO$_3$).

$2 NO_2 (g) + H_2O (l) \rightarrow$

$HNO_2 (aq) + HNO_3 (aq)$

Like with sulfuric acid, these acids produce acid rain when mixed with precipitation.

Chapter 7 Exercises

1. Define a Bronsted-Lowery acid

   _____

   _____

   _____

2. Define a Arrhenius acid and base

   _____

   _____

   _____

3. Which is a basic or alkaline substance?

   a. gastric fluid   b. black coffee

   c. vitamin C   d. oven cleaner

4. Which is an acidic substance?

   a. household bleach

   b. apples

   c. vinegar

   d. oven cleaner

5. Which of the following is a property of bases?

   a. feel slippery to the touch

   b. have pH below 7

   c. turn blue litmus red

   d. neutralize substances like NaOH

6. Define a strong acid:

_____

_____

_____

_____

7. Which of the following is not strong acid?

   a. HNO_3 b. H_3PO_4 c. HCl d. HBr

8. Which equation represents a neutralization reaction?

   a. $KCl + H_2O \rightarrow K^+(aq) + Cl^-(aq)$

   b. $H_3O^+(aq) + OH^-(aq) \rightarrow H_2O$

   c. $HCl + H_2O \rightarrow H_3O^+(aq) + Cl^-(aq)$

   d. $NaOH + H_2O \rightarrow Na^+(aq) + OH^-(aq)$

9. The strength of an acid is related to its

   a. extent of ionization

   b. reaction with a salt

   c. concentration

   d. commercial ranking in the economy

10. Which of the following indicates a basic solution?

    a. $pH = 7$

    b. $pH < 7$

    c. $pH = 11$

    d. $pH = 0$

11. In an acidic solution,

   a. $[H_3O^+]$ is greater than $[OH^-]$

   b. $[H_3O^+]$ equals $[OH^-]$

   c. $[OH^-]$ is greater than $[H_3O^+]$

   d. none of these

12. Which common substance would have a pH less than 7?

   a. oven cleaner

   b. wine

   c. ammonia

   d. bleach

13. A buffer is a mixture that

   a. maintains pH

   b. causes a solution not to conduct electricity

   c. neutralize salts

   d. causes high blood pressure

14. A solution with pH = 4.00 has

   a. relatively high concentration of OH^-

   b. relatively low concentration of H_3O^+

   c. zero concentration of OH^-

   d. relatively high concentration of H_3O^+

15. What is the pH of a 0.0001 M HCl solution?

   _____

16. What is the pH of a 0.035 M HCl solution?

   _____

17. What is the salt formed when an HCl solution reacts with $Mg(OH)_2$?

   a. $MgCl_2$

   b. Mg_2Cl

   c. $MgCl$

   d. Mg_2Cl_2

18. The substance $Ca(OH)_2$ is

   a. an acid

   b. a hydrate

   c. a base

   d. an oxide

19. Red cabbage can be used as a dye indicator used to measure pH, in basic solutions it has a _____ color.

20. H_3O^+ is the

   a. hydronium ion

   b. hydrogen ion

   c. proton

   d. hydridium ion

21. What is the pH for a solution with hydrogen ion molarity of 0.01?

_____

22. If 3.00 moles of a substance are dissolved in 500 mL (0.5 L) of solution, the molarity of this solution is

_____M

23. The symbol, M, related to concentration of solution, refers to

a. mass

b. molecular weight

c. moles of solute dissolved in a liter of solution

d. none of these

24. What is the molarity of a solution prepared by dissolving 1.50 moles of lithium hydroxide in enough water to make 400 ml of solution?

_____M

25. What is the molarity of a solution prepared by dissolving 1.5 grams of lithium hydroxide in enough water to make 400 ml of solution?

_____M

26. What is the concentration of the hydronium ion in a 1.5 L solution containing 1.5 grams of HCl?

_____M H_3O^+

27. What is the pH of a 1.5 L solution containing 1.5 grams of HCl?

_____

28. A base is

    a. an OH$^-$ ion donor

    b. a hydrogen ion donor

    c. a substance like magnesium hydroxide, $Mg(OH)_2$

    d. both a and c

29. Why is pure water neutral?

    _____

    _____

    _____

    _____

30. Which is a weak acid?

    a. HCl

    b. H_2SO_4

    c. HNO_3

    d. $HC_2H_3O_2$

31. An solution of an acid ion pair such as H_2CO_3 and HCO_3^- qualifies as a

    a. strong acid-strong base pair

    b. buffer system

    c. substitute for hemoglobin

    d. acidic solution

32. Chemical buffering systems

    a. maintain constant pH

    b. consist of a conjugate acid-base pair

    c. absorb added H^+ or OH^- ions

    d. do all of the above

33. Which of the following is true about a solution with pH = 5.00?

    a. the solution is basic

    b. the OH^- concentration > H^+ concentration

    c. the solution is acidic

    d. both a and b

34. Which of the following is true about a solution with pH = 8.00?

    a. the solution is basic

    b. the H^+ concentration equals 0.00000001 M

    c. the OH^- concentration > H^+ concentration

    d. all of these

35. Acid rain is defined as any precipitation with a pH less than:

    a. 7.00   b. 5.60   c. 8.00   d. 12,56

36. Which produces more NO and NO_2 in the U.S.

    a. electric utilities

    b. inductrial sources

    c. transportation

    d. all of these produce the same

Chapter 8

Chemistry of Carbon "The Organics"

With a 109 different elements in the periodic table, you might ask why *carbon* deserves an entire chapter of this book. The answer lies with the number and importance of carbon containing compounds. There are 50 times the numbers of carbon containing compounds than all the other elements combined, with a current estimate of well over 10 million. Carbon is also unique in that it has the ability to bond to itself (a process called **catenation**), as well as to other elements allowing it to make molecules that range in size from very small to very large. For instance, methane (CH_4), the smallest carbon compound, has a mass of 18.0 g/mol, whereas proteins, carbon containing biomolecules, can have masses higher than 1,000,000 g/mol. Carbon therefore, not only gets a dedicated chapter but an entire field of study called organic chemistry. **Organic chemistry** is the study of carbon and the compounds it makes called **organic compounds**; all compounds made from other elements are called **inorganic compounds**. Our bodies are composed of organic compounds such as carbohydrates, proteins, fats, and other carbon containing biomolecules (Chapter 9). Organic chemistry got its name from the idea that compounds produced from living organisms contained a so-called *life-force* and therefore these compounds could not be produced in a laboratory. However, in 1828 a German chemist, Frederick Wohler, determined that it was indeed possible to synthesize organic compounds from those compounds that were considered inorganic. One of the first organic compound synthesized was urea, which is a by product of urine metabolism. Since then millions of organic compounds have been synthesized. Organic chemistry is therefore, the largest and fastest growing branch of chemistry. Because of the shear numbers of new organic molecules produced each year, a classification system was introduced to separate these organic compounds into *families*. Each **organic family** consists of compounds that have a chemically active center called a **functional group** and so they have very similar chemical properties. In this chapter we will limit our attention to nine of the organic families to include:

- Alkanes
- Alkenes
- Alkynes
- Aromatics
- Alcohols
- Ethers
- Aldehydes
- Ketones
- Carboxylic Acids

We will also discuss the properties of some of these compounds as they relate to energy sources and their uses in the manufacturing of consumer products. The following key questions will also be addressed:

What are the major types of hydrocarbons?
What are isomers?
What are functional groups?
What are the names of organic compounds?
What are hydrocarbon fuels and how are they produced?

141

Organic Families

Of the nine families of organic compounds mentioned in the introduction, alkanes, alkenes, alkynes, and aromatics fall under a separate class called hydrocarbons. **Hydrocarbons** are organic compounds containing atoms of carbon and hydrogen only.

Alkanes

Alkanes are organic compounds made entirely of carbon-carbon and carbon-hydrogen single bonds. Because they contain only single bonds, alkanes are also called **saturated hydrocarbons**. Alkenes, alkynes and, aromatics are compounds with carbon-carbon multiple bonds and are, therefore called **unsaturated hydrocarbons**. Alkanes have the general molecular formula:

$$C_nH_{2n+2}$$

Where n = the number of carbon atoms.

For example, for n = 3 (an alkane with three carbons) the molecular formula would be:

$$C_3H_8$$

Since all members of the alkane family have the same ratios of carbon and hydrogen atoms a list of all saturated hydrocarbons could be easily ascertained. Table 8-1 lists the first ten un-branched alkanes along with their molecular formulas, condensed structural formulas, names, and melting and boiling points. **Condensed structural formulas** are an abbreviated or shorthand representations of the order in which the carbon atoms are bonded together in hydrocarbons. The hydrocarbons given in the table are shown to be bonded in a continuous straight order without any branching of the carbon atoms. **Expanded structural formulas** use single lines to show every carbon-carbon as well as every carbon-hydrogen bond in the molecule.

n	Molecular Formula	Condensed structural formula	Name	Melting point ^{o}C	Boiling point ^{o}C
1	CH_4	CH_4	methane	-182	-162
2	C_2H_6	CH_3CH_3	ethane	-183	-89
3	C_3H_8	$CH_3CH_2CH_3$	propane	-190	-42
4	C_4H_{10}	$CH_3CH_2CH_2CH_3$	butane	-138	-1
5	C_5H_{12}	$CH_3CH_2CH_2CH_2CH_3$	pentane	-130	36
6	C_6H_{14}	$CH_3CH_2CH_2CH_2CH_2CH_3$	hexane	-95	69
7	C_7H_{16}	$CH_3CH_2CH_2CH_2CH_2CH_2CH_3$	heptane	-91	98
8	C_8H_{18}	$CH_3CH_2CH_2CH_2CH_2CH_2CH_2CH_3$	octane	-57	126
9	C_9H_{20}	$CH_3CH_2CH_2CH_2CH_2CH_2CH_2CH_2CH_3$	nonane	-51	151
10	$C_{10}H_{22}$	$CH_3CH_2CH_2CH_2CH_2CH_2CH_2CH_2CH_2CH_3$	decane	-30	174

Table 8-1
Formulas and properties of first ten alkanes

For example, propane (C_3H_8) has the condensed structural formula:

CH₃ CH₂CH₃

This would translate to an expanded structural formula as:

```
    H   H   H
    |   |   |
H — C — C — C — H
    |   |   |
    H   H   H
```

Propane (expanded structural formula)

Example 8-1

Draw the expanded structural formula for pentane.

Answer

Pentane has the molecular formula C_5H_{12}, and the condensed molecular formula of:

CH₃CH₂CH₂CH₂CH₃

The expanded structural formula would show five carbon atoms bonded in a straight line and each connected by a single bond. The hydrogen's attached to each carbon would then be drawn using a single line.

```
    H   H   H   H   H
    |   |   |   |   |
H — C — C — C — C — C — H
    |   |   |   |   |
    H   H   H   H   H
```

Expanded structure of pentane

Structural Isomers

Butane and 2-methylpropane have the same molecular formula, C_4H_8, but are very different compounds with different structures and properties. These compounds are examples of **isomers**, different compounds with the same molecular formula. There are several different types of isomers. Butane and 2-methylpropane are examples of **structural isomers**- compounds that differ from each other in the order in which the atoms are bonded.

The expanded structural formulas of butane and 2-methylpropane are shown in Figure 8-2.

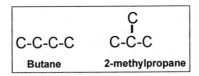

Figure 8-1
Expanded structural formulas of the isomers butane and 2-methyl propane; the hydrogen's have been removed for clarity but each carbon has enough hydrogen's making four bonds to each carbon.

Comparing the two isomers in Figure 8-1, we see the basic difference in their structures is the number of carbon atoms connected in a successive manner. Butane has four successive carbons, while 2-methylpropane has only three. These numbers represent the longest continuous carbon chains, or simply the **longest chain** found in each hydrocarbon.

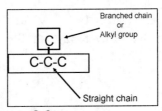

Figure 8-2
Branched chain hydrocarbon

Butane represents an un-branched hydrocarbon while 2-methylpropane represents a branched hydrocarbon. The groups attached to the longest chain of a branched hydrocarbon are called **alkyl groups** (Figure 8-2). The names of alkyl groups are based on the number of carbon atoms they contain and are derived from those of the ten straight-chained hydrocarbons (parent compounds). The ending of the name of the parent compound is then changed from *–ane* to *–yl*. For example, an alkyl group consisting of one carbon would use the parent name *methane*, and would be called *methyl* (Figure 8-3).

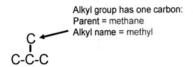

Condensed formula CH₃CH(CH₃)CH₃

Figure 8-3
   Example of naming alkyl groups

Notice that in Figure 8-3, the alkyl group is placed inside a set of parentheses in the condensed structural formula, (CH3), and follows the carbon to which it is attached.

Naming Branched Chained Hydrocarbons

The **IUPAC** (international union of pure and applied chemist) provides a set of rules for naming alkanes. If the IUPAC rules are followed correctly, they will always result in the same name for the same compound. These rules are as follows:

1. The name of the longest chain becomes the parent name of the compound (failure to correctly name the parent name will result in an incorrect name even if the subsequent rules are followed correctly).
2. The parent name only accounts for the carbons in the longest chain. The carbons that are not part of the longest chain (alkyl groups) must also be included in the name of the compound.

   a) The name(s) of the alkyl groups is placed in front of the parent name.

   b) Use the prefixes di-, tri- or tetra- before the name of the alkyl group when there are two, three, or four alkyl groups with the same name.

   c) Name the groups in alphabetical order when there are two or more different types of groups, ignoring all prefixes such as d-, tri- etc.

3. Number the carbons in the longest chain, starting from which ever end results in the lowest number (or set of lowest numbers) for the alkyl group(s).
4. In front of the name of each alkyl group, place the number of the carbon to which the group is attached.
5. Use hyphens to separate numbers from words; use commas to separate numbers.

Example 8-2

Give the IUPAC name for the following compound:

CH3CH(CH3)CH2CH2CH3

Answer

From the condensed structural formula we can expanded it to:

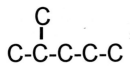

Notice that the hydrogen's have been removed for clarity. Then using the IUPAC rules we can name it as follows:

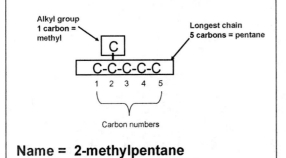

Name = 2-methylpentane

Example 8-3

Give the IUPAC name for the following compound:

CH$_3$CH$_2$CH$_2$CH(CH$_2$CH$_3$)C(CH$_3$)$_2$CH$_2$CH$_3$

Answer

From the condensed structural formula we can expanded it to:

Using IUPAC we get

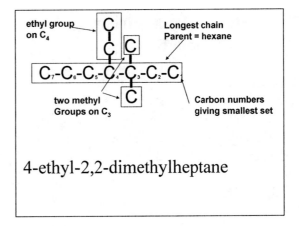

4-ethyl-2,2-dimethylheptane

Properties of Alkanes

Note in Table 8-1 that the first four hydrocarbons, methane, ethane, propane, and butane, are all gases at room temperature (25 °C); their boiling points are less than 0 °C, while the remaining six are liquids. Hydrocarbons with eighteen or more carbon atoms are solids at room temperature. For instance, octadecane (C$_{18}$H$_{38}$) melts at 28 °C and boils at 343 °C. Hydrocarbons are non-polar molecules that are essentially insoluble in water. When a liquid hydrocarbon is mixed with water, two layers are formed- the top layer being the hydrocarbon since its density is much less than that of water. They dissolve in many organic substances of similar polarities such as fats, oils, and waxes. Alkanes are limited to only a few chemical reactions, with the most important one being combustion. During the combustion of alkanes, a large amount of heat is released which can be used to do work. For this reason, alkanes are used mainly as fuels.

Hydrocarbons with Carbon-Carbon Multiple Bonds

Alkenes, alkynes, and aromatic are a class of compounds containing carbon-carbon multiple bonds and are called unsaturated hydrocarbons. These groups of hydrocarbons, like alkanes, are less dense than water, insoluble in water, and combustible.

Alkenes

Alkenes are hydrocarbons containing one or more carbon-carbon double bonds. Alkenes have the general formula:

$$C_nH_{2n}$$

For example, for n = 3 (an alkene with three carbons) the molecular formula would be:

$$C_3H_6$$

Naming Alkenes

The IUPAC rules for naming alkenes are very similar of those for alkanes but with two important differences. First, the parent chain must include the double bond even if it is not the longest. Second, the parent chain must be numbered as to give the location of the double bond the smaller value. This overrides the numbering of the alkyl groups. The location number of the double bond, followed by a hyphen, precedes the parent name. The parent name of the alkene derives from the name of the alkane with the same number of carbons, but the ending is changed from –*ane* to –*ene*.

Example 8-4

Give the IUPAC name for the following compound:

$CH_3CH(CH_3)CH_2CH=CH_2$

Answer

From the condensed structural formula we can expanded it to:

```
      C
      |
C-C-C-C=C
```

Notice that the hydrogen's have been removed for clarity. Then using the IUPAC rules we can name it as follows:

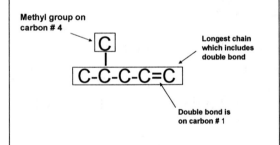

4-methyl-1-pentene

Structural Isomers of Alkenes

1-butene and 2-butene have the same molecular formula (C_4H_8) but are quite different compounds. Like alkanes, 1-butene and 2-butene represent structural isomers. The difference is in the location of the double bond, as shown in Figure 8-4.

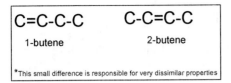

Figure 8-4
Structural isomers of butene

Alkynes

Alkynes are unsaturated hydrocarbons containing at least one carbon-carbon triple bond and have the general formula:

$$C_nH_{2n-2}$$

For example, for n = 3 (an alkene with three carbons) the molecular formula would be:

$$C_3H_4$$

Naming Alkynes

Alkynes are named just like alkenes, but the ending of the parent hydrocarbon is changed to –*yne*.

Example 8-5

Give the IUPAC name for the following compound:

CH₃CH(CH₃)CH₂C≡CH

Answer

From the condensed structural formula we can expanded it to:

```
    C
    |
C-C-C-C≡C
```

Notice that the hydrogen's have been removed for clarity. Then using the IUPAC rules we can name it as follows:

Methyl group on carbon # 4
Longest chain which includes triple bond
triple bond is on carbon # 1

4-methyl-1-pentyne

Aromatic Hydrocarbons

An **aromatic compound** is any compound that contains the benzene ring or has properties similar to the benzene ring. The structure of benzene consists of six carbon atoms bonded to each other in such a way as to form a ring with alternating double and single bonds between the carbon atoms and is represented by two forms as shown in Figure 8-5.

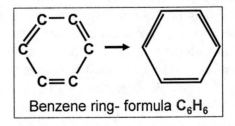

Benzene ring- formula C₆H₆

Figure 8-5
Two representations for benzene

Aromatic hydrocarbons are generally described as having a very pleasant odor. However, they are also known to cause serious illnesses like cancer and leukemia after long exposures. They are primarily used as solvents and fuels. One benzene derivative you are probably familiar with is naphthalene. This white crystalline solid, used as an insecticide, gives the characteristic odor of mothballs.

Polymers

Polymers are substances with very high molar masses and are composed of a large number of repeating units. Polymers occur both naturally and are synthetically prepared. Naturally occurring polymers include proteins, starches, cellulose, and latex. Synthetic polymers are produced commercially on a very large scale and have a wide range of properties and uses. For instance plastics are all synthetic polymers.

Polymers are formed by chemical reactions in which a large number of smaller molecules, or units, called monomers are joined together. Most polymers contain only one type of monomer; however, two or three different monomers may be combined to form polymers.

Polymers are classified by the reactions by which they were formed. If all atoms in the monomers are incorporated into the polymer, the polymer is called **an addition polymer**. If some of the atoms of the monomers are released into small molecules, such as water, the polymer is called a **condensation polymer.** Addition polymers are made from monomers containing a double bond between carbon atoms, like alkenes. Since these monomers are used to make polymers, they are given a special name called **olefins**. Table 8-2 lists some common addition polymers and the monomers used in their production.

Monomer	Polymer	Uses
Ethylene	Polyethylene	Plastics, films, toys
Propylene	Polypropylene	Indoor/ outdoor Carpet
Vinyl chloride	PVC (polyvinyl chloride)	Plastic pipes rain coats
Styrene	Polystyrene	Food and drink coolers

Table 8-2
Common addition polymers

Hydrocarbons Containing Oxygen

Alcohols, ethers, aldehydes, ketones, and carboxylic acids are a class of hydrocarbons containing oxygen. Alcohols and ethers are grouped together and referred to as organic derivatives of water. Aldehydes, ketones, and carboxylic acids are associated as compounds containing the carbonyl group (C=O). These organic compounds are also produced from the oxidation of alcohols.

Alcohols and Ethers

An alcohol is any compound with an (OH) functional group attached to a carbon atom, which is also attached to three other groups by single bonds, as shown in Figure 8-6.

148

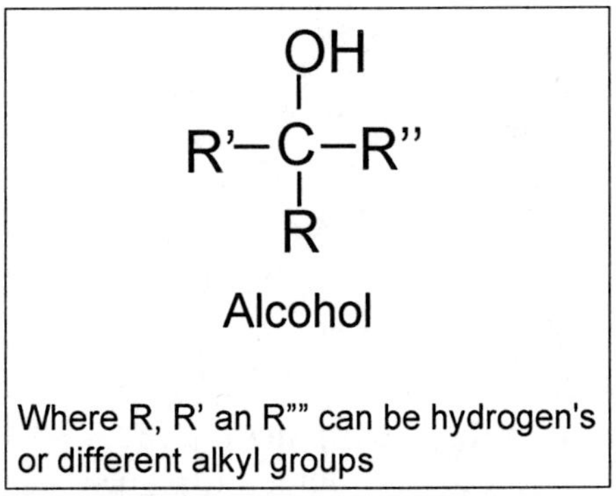

OH
|
R'—C—R"
|
R

Alcohol

Where R, R' an R"" can be hydrogen's or different alkyl groups

Figure 8-6

Structure of an alcohol

There are three classes of alcohols, primary, secondary, and tertiary, based on the type of carbon to which the (OH) group is attached. A primary alcohol is one in which the (OH) group is attached to a carbon atom that is bonded to only one other carbon atom. In a secondary alcohol, the (OH) group is attached to a carbon that is bonded to two other carbons. As for a tertiary alcohol the (OH) carbon is bonded to three other carbon atoms (Figure 8-7)

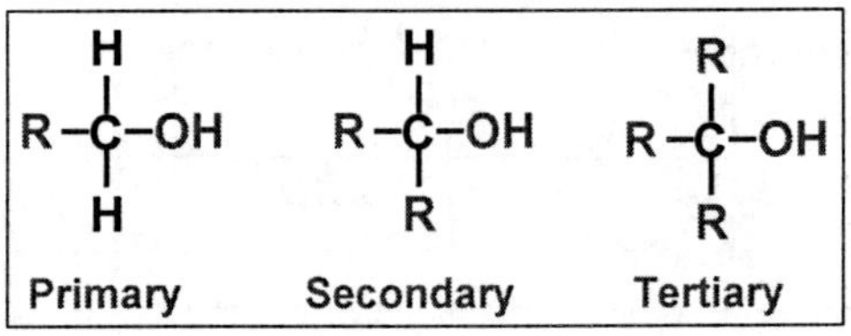

H   H   R
| | |
R—C—OH R—C—OH R—C—OH
| | |
H   R   R

Primary   **Secondary**   **Tertiary**

Figure 8-7

Primary, secondary and tertiary alcohol

Alcohols are named by removing the (–e) from the end of the parent hydrocarbon and replacing it with (–ol). For instance, the simplest alcohol is derived from methane (CH_4) and is called methanol (CH_3OH) (also known as wood alcohol). Methanol is highly toxic, causing many people who consumed it, to become blind or die. Ethanol is the al-

cohol associated with "alcoholic" beverages, and has been around for at least 6000 years. It is produced by the fermentation of yeast in solutions of either sugars or starches. Ethanol is not as toxic as methanol, but it is still dangerous. Most people are intoxicated at blood alcohol levels of about 0.1 gram per 100 mL. An increase in the level of alcohol in the blood to between 0.4 and 0.6 g/100 mL can lead to coma or death.

Properties of Alcohols

As was mentioned earlier, hydrocarbons are insoluble in water. However, some alcohols are quite soluble. For instance, methanol, ethanol, and propanol are completely soluble in water in all proportions. This solubility is due to the strong hydrogen bonding between water and the (OH) functional group. However, as the carbon chain becomes longer, the solubility of alcohols in water decreases, as shown in Table 8-3. Alcohols can be thought of as having two ends; one contains a very polar (OH) group and the other a nonpolar hydrocarbon group. The end containing the polar (OH) group that can form hydrogen bonds to neighboring water molecules and is, therefore, said to be "water loving" or **hydrophilic**. While the hydrocarbon end is so nonpolar it is said to be "water hating" or **hydrophobic**. As the hydrocarbon chain becomes longer, the hydrophobic end of the molecule increases, and the solubility of the alcohol in water gradually decreases until it becomes essentially insoluble in water.

149

Formula	Name	Solubility in water (g/100g)
CH_3OH	methanol	Completely
CH_3CH_2OH	ethanol	Completely
$CH_3(CH_2)_2OH$	propanol	Completely
$CH_3(CH_2)_3OH$	butanol	9.0
$CH_3(CH_2)_6OH$	heptanol	0.18
$CH_3(CH_2)_9OH$	decanol	insoluble

Table 8-3

Solubilities of alcohols in water

The use of alcohols as alternative fuel sources, either alone or in combination with other fuels, has been given a lot of attention lately because of its possible replacement for the dwindling availability of fossil fuels. Both ethanol and methanol have been considered for this purpose. While both can be obtained from petroleum or natural gas, ethanol can be produced from sugar cane or corn and is, therefore, considered a renewable resource. When alcohol is mixed with fossil fuels, it is called **gasohol** (a mixture of 85% corn alcohol (ethanol) and 15% gasoline), which are currently available at gas stations around the United States. Table 8-4 lists some important alcohols and their uses.

Formula	BP (°C)	Synthetic name	Common name	use
CH_3OH	65	Methanol	Methyl alcohol	Fuel additives
CH_3CH_2OH	76	Ethanol	Ethyl alcohol	Beverages, fuel additives, solvents
$CH_3CH_2CH_2OH$	97	1-propanol	Propyl alcohol	Solvents
$CH_3CH(OH)CH_3$	82	2-propanol	Isopropyl alcohol	Rubbing alcohol
$CH_2(OH)CH_2(OH)$	198	1,2-ethanediol	Ethylene glycol	Antifreeze
$CH_2(OH)CH_2(OH)\ CH_2(OH)$	290	1,2,3-propanetriol	Glycerol	Moisturizers

Table 8-4

Important alcohols and their uses

Oxidation Products of Alcohols

When primary alcohols are oxidized by the addition of oxygen, two products are formed--an aldehyde and a carboxylic acid (Figure 8-8)

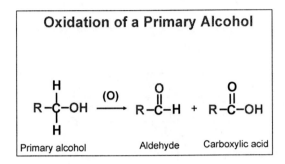

Figure 8-8
Oxidation of a primary alcohol

The oxidation of a secondary alcohol results in the formation of a ketone see Figure 8-9.

Figure 8-9
Oxidation of a secondary alcohol

Table 8-5 lists some common aldehydes, ketones and carboxylic acids. Table 8-6 lists some common carboxylic acids and their common names.

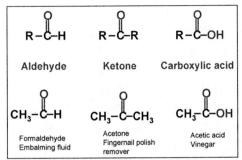

Table 8-5
Common oxidation products from alcohols

Formula	Common name	Synthetic name (Acid)	Where found
HCOOH	Formic acid	Methanoic	Ant stings
CH$_3$COOH	Acetic acid	Ethanoic	Vinegar
CH$_3$CH$_2$COOH	Propionic acid	Propanoic	Swiss cheese
CH$_3$(CH$_2$)$_2$COOH	Butyric Acid	Butanoic	Rancid butter
CH$_3$(CH$_2$)$_3$COOH	Valeric Acid	Pentanoic	Cow manure

Table 8-6
Common carboxylic acids and their common names

Ethers

Ethers are compounds containing an oxygen atom bonded between to alkyl groups as shown in Figure 8-9.

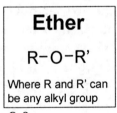

Figure 8-9
Structure of ether

The most commonly used ether, diethyl ether, has two ethyl groups on either side of the oxygen. Diethyl ether is primarily used as a solvent because it dissolves many organic substances that are insoluble in water. Diethyl ether is highly flammable and must be handled with

151

extreme care. Also, they tend to react over time with oxygen, forming peroxides, which are very explosive.

Diethyl ether was introduced as an anesthetic in 1842 and became the most widely used form of general anesthetic for surgery. However, today it is no longer used on humans because of its undesirable side effects.

Energy from Hydrocarbons

There are many hydrocarbons that are used extensively as a source of energy. The gasoline in your car, the coal that is used to generate electricity, and the natural gas used to heat your home are just a few examples of the usefulness of some hydrocarbons as a source of energy. Hydrocarbons that are used to produce energy are called fuels. **Fuels** are reduced forms of matter which burn easily in the presence of oxygen, producing large quantity of heat which can be used to perform work. This energy (heat), stored in the chemical bonds of the fuels, is released during the oxidation reaction (combustion) of the fuels. For example, when one mole of natural gas (methane) is burned, each carbon-hydrogen bond is broken, as well as the oxygen-oxygen bonds. These free atoms then reform, producing carbon dioxide and water, releasing 210,000 calories of heat (Figure 8-10). The amount of energy released not only depends on the amount of fuel used, but on the type of fuel used. Different hydrocarbons release different amounts of energy when equal masses are combusted. The **energy value of a fuel** is the amount of energy released per gram combusted. For example, hydrogen produces three times as much energy as gasoline when equal amounts are combusted.

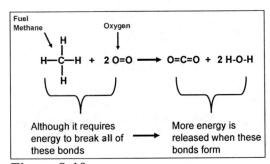

Figure 8-10
Combustion of natural gas

Table 8-7 gives the energy values for some common fuels.

Substance	Heat (kcal/g)
Hydrogen	34.0
Natural gas (methane)	11.7
Gasoline	11.5
Coal	7.4
wood	4.3

Table 8-7
Energy values for some common fuels.

Fossil fuels are complex mixtures of hydrocarbons which can be separated into three major forms of fuels--petroleum, natural gas, and coal. 86% of the energy produced in the United States is generated from fossil fuels (Figure 8-11).

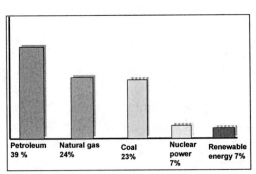

Figure 8-11
Sources of energy in the United States

Of the remaining 14% of the energy produced, 7% comes from nuclear power and 7% is generated from natural resources, as shown in Table 8-8.

Energy source	% Usage	Energy source	% usage
Solar	1.0 %	Waste	8.0 %
Wind	1.0 %	Wood	38.0 %
Alcohol	2.0 %	Hydroelectric	46.0 %
Geothermal	5.0 %		

Table 8-8
Breakdown of renewable energy in the United States

Refining of Petroleum

Refining is the process of separating the thousands of hydrocarbons present in crude petroleum, one of which is gasoline. This process is based on the principle that the longer the carbon chain, the higher the temperature at which the compounds will boil. The crude oil is changed into a gas by heating. The gases are passed through a distillation column which becomes cooler as the height of the column increases. When a gaseous compound cools below its boiling point, it condenses into a liquid and may be removed from the column. Since all the fractions have different boiling points they can all be separated by this method. Although all of the fractions collected are useful, the greatest demand is for gasoline. Gasoline, however, only makes up about 35% of a barrel of crude oil. The demand for this fuel requires that over 50% of the crude oil be converted into gasoline. To meet this demand some petroleum fractions must be converted to gasoline. This is accomplished by **catalytic cracking,** a process by which larger kerosene fractions are broken down and converted into hydrocarbons in the gasoline range. This process involves heating saturated hydrocarbons, having a carbon chain length

from twelve to sixteen, under high pressure in the absence of oxygen. These hydrocarbons literally break into smaller hydrocarbons with chain lengths between five and twelve carbons; the range of carbons found in "straight run" gasolines. Straight run gasolines primarily contain straight-chain hydrocarbons, which burn too rapidly and cause uncontrolled explosions of the fuel; these explosions are characterized by a "knocking" or a "pinging" sound in the engine. The knocking is drastically reduced when the fuel contains branched-chained hydrocarbons. **Octane rating** is an arbitrary scale for rating the relative knocking tendencies of gasolines. Heptane (C_7H_{16}), a straight-chain hydrocarbon, has been assigned an octane rating of 0 because it knocks considerably when used as auto fuel. On the other hand, 2,24-trimethylpentane ($CH_3C(CH_3)_2CH_2CH(CH_3)CH_3$) burns with little or no knocking and is assigned an octane rating of 100. All other gasolines are then measured by this scale. **Catalytic re-forming** is a process used to increase the octane rating of straight-run gasolines by converting straight-chain hydrocarbons to branched-chain hydrocarbons.

Besides gasolines there are a large number of useful organic chemicals obtained from fossil fuels by catalytic cracking and fractional distillation techniques. Figure 8-12 shows the major products obtained from the catalytic cracking of alkanes and the materials generated from them.

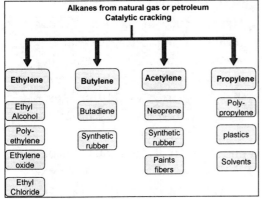

Figure 8-12
Products from catalytic cracking of alkanes and the materials generated from them

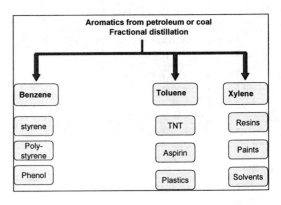

Figure 8-13
Products from fractional distillation of petroleum and the materials generated from them

Table 8-9 lists several organic materials produced in the U.S., their methods of production and uses.

Name	Produced	Uses
Ethylene	Cracking oil and natural gas	Plastics, fibers and solvents
Propylene	Cracking oil and oil products	Plastics, fibers and solvents
Urea	Reaction NH_3 and CO_2 under pressure	Fertilizer, animal feeds, adhesives
Styrene	Dehydration of ethylbenzene	Polymers, polyester

Table 8-9
Organic materials produced in the U.S

Figure 8-13 shows the major aromatics produced from fractional distillation of petroleum and the materials they generate.

Chapter 8 Exercises

1. Which is a characteristic feature of alkenes?

    a. general formula of C_nH_{2n+2}

    b. tetrahedral geometry

    c. unsaturated

    d. hydrocarbons which contain only single bonds

2. Which hydrocarbon is the principal component of natural gas?

    a. methane

    b. benzene

    c. ethylene

    d. isooctane

3. Which is a characteristic of alkanes?

    a. contain C=C

    b. general formula C_nH_{2n}

    c. saturated

    d. none of these

5. Which fossil fuel furnishes the most heat energy per gram?

    a. petroleum

    b. charcoal

    c. natural gas

    d. coal

6. All combustion reactions of fossil fuels

   a. give off energy

   b. are complete combustions with oxygen

   c. are endothermic reactions

   d. are 100% efficient

7. Butane [$CH_3CH_2CH_2CH_3$] and 2-methylpropane [$(CH_3)_2CHCH_3$] are

   a. same compound   b. alkenes   c. structural isomers   d. alkynes

8. Which of the following will not improve the octane rating of a gasoline?

   a. increasing the percentage of branched-chain hydrocarbons

   b. increasing the percentage of aromatic hydrocarbons

   c. increasing the percentage of straight-chain hydrocarbons

   d. adding octane enhancers

9. Which process is used to increase the amount of the gasoline fraction in refining to match commercial demand?

   a. reforming   b. oxygenating   c. cracking   d. gasification

10. A compound containing seven carbon atoms is

    a. heptane   b. octane   c. hexane   d. nonane

11. The name of the C_2H_5- group is

    a. methyl   b. ethyl   c. propyl   d. butyl

12. What is the name of this compound?

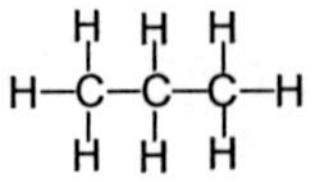

    _____

13. The complete combustion of C_2H_6 yields

    a. C_2H_4 and H_2

    b. CO and H_2O

    c. CO_2 and H_2O

    d. C and H_2

14. Hydrocarbons with triple bonds are called

    a. alkanes   b. alkenes   c. alkynes   d. aromatic hydrocarbons

15. Which of the following is ethene?

    a. $CH_2=CHCH_3$

    b. $CH_2= CH_2$

    c. H-C-C-H

    d. CH_3-CH_3

16. Which alcohol is called wood alcohol?

    a. ethanol

    b. isopropyl alcohol

    c. 1-propanol

    d. methanol

17. Gasohol contains

    a. up to 85% ethanol mixed with unleaded gasoline

    b. up to 10% methanol mixed with unleaded gasoline

    c. three grams of tetraethyl lead

    d. three grams of tertiary-butyl alcohol per gallon

18. Gasoline is a

    a. single, simple hydrocarbon

    b. single, complex hydrocarbon

    c. mixture of C_{12}-C_{16} hydrocarbons

    d. mixture of C_5-C_{12} hydrocarbons

19. Which of the following has an octane rating of 0

    a. heptane

    b. isooctane

    c. hexane

    d. nonane

20. Which of the following compounds would you expect to have the highest octane rating?

    a. pentane, C_5H_{12}, $CH_3CH_2CH_2CH_2CH_3$

    b. hexane, C_6H_{14}, $CH_3CH_2CH_2CH_2CH_2CH_3$

    c. isooctane, C_8H_{18}, $(CH_3)_3CCH_2CH(CH_3)_2$

    d. heptane, C_7H_{16}, $CH_3CH_2CH_2CH_2CH_2CH_2CH_3$

21. The octane rating of straight run gasoline from fractional distillation of petroleum can be raised by

    a. catalytic reforming

    b. adding tertiary-butyl alcohol

    c. adding ethanol

    d. all of the above

158

22. Fractional distillation of petroleum

   a. separates and purifies all of the hydrocarbons in crude oil

   b. separates fractions, each of which contain many hydrocarbons

   c. is the final step in refining gasoline

   d. fractions large molecules into smaller ones

23. Name the following compounds

a. 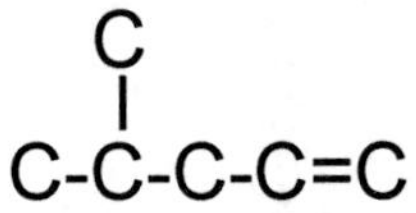
$$C-C-C-C=C$$ with a C branch on the third carbon

b.
$$C-C-C-C-C$$ with a C branch on the second carbon

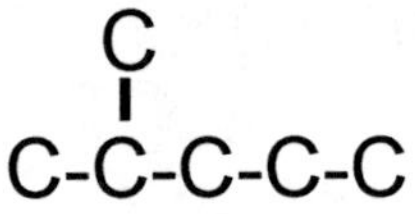

c.
$$C-C-C-C-C$$ with C branches above and below the second carbon

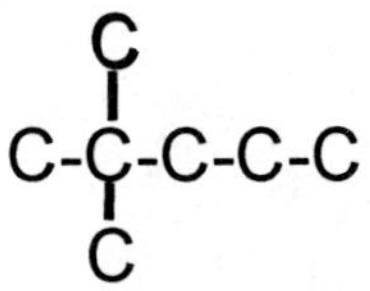

a _____

b _____

c _____

24. Which fossil fuel burns cleaner and more completely?

   a. coal   b. natural gas   c. petroleum   d. wood

25. Which is a characteristic of alkanes?

   a. contain carbon carbon triple bonds C:::C

   b. general formula C_nH_{2n-2}

   c. have only single bonds

   d. none of these

26. Which of the following does not contain oxygen?

    a. alkanes

    b. alcohols

    c. ketones

    d. carboxylic acids

27. An organic compound with an OH functional group is a(n)

    a. aldehyde   b. ketone   c. ether   d. alcohol

28. The oxidation of primary alcohols produces

    a. carboxylic acids   b. amines   c. ketones   d. esters

29. Which carboxylic acid is found in vinegar?

    a. butyric   b. propionic   c. acetic   d. formic

30. The small molecules used to synthesize polymers are called

    a. macromolecules   b. monomers   c. elastomers   d. peptides

31. Which is not an addition polymer?

    a. PVC   b. latex   c. polyethylene   d. polystyrene

32. An addition polymer that finds use in outdoor carpets

    a. polypropylene   b. PVC   c. nylon   d. polyacrylonitrile

33. Which polymer material occurs naturally?

    a. nylon   b. dacron   c. polycarbonates   d. latex rubber

34. Polymerization is a(n)

    a. formation of large molecules out of small ones

    b. formation of small molecules out of large ones

    c. producing gasoline

    d. means of producing molecules of intermediate size, but not very large ones

35. Which of the following is an incorrect pairing?

    a. $CH_2=CHCH_2CH_2CH_2CH_2CH_2CH_3$, alkene

    b. CH_3OCH_3, aldehyde

    c. CH_3CH_2OH, alcohol

    d. $(CH_3)_2CO$, ketone

Unit Conversions

======➡ ⬅======	Multiply by Divide by	======➡ ⬅======
Length		
Meter (m)	39.37	Inch (in)
Meter (m)	3.281	Feet (ft)
Millimeter (mm)	0.001	Meter (m)
Centimeter	0.3937	Inch (in)
Foot (ft)	0.305	Meter (m)
Pressure		
Atmosphere	4.697	Pounds per in^2
Mass		
Kilogram (kg)	2.2	Pound (lb)
Ounce (oz)	28.35	Gram (g)
Volume		
Cubic centimeter(cm^3)	0.061	Cubic inch (in^3)
Cubic feet (ft^3)	0.028	Cubic meter (cm^3)
U.S. gallon	3.785	Liter (L)
U.S. gallon	0.1337	Cubic feet (ft^3)
Ounce (U.S. fluid)	29.57	Cubic centimeter (cm^3)

164

Laboratory Manual and Laboratory completion form

Upon completion of each experiment you must have the laboratory teaching assistant sign your completion form. Failure to have it signed could result in a lower lab grade.

TA's Signature

Experiment 1&2 _____

Experiment 3 _____

Experiment 4 _____

Experiment 5 _____

Experiment 6&7 _____

Experiment 8 _____

166

Experiment

1

Safety and Laboratory Techniques

Objective

To become familiar with general safety rules, working safely with chemicals and hazard identifications in the laboratory. You will also learn to use some of the equipment found in the laboratory.

General Safety Rules

The following rules must be observed while working in the laboratory.

1. **Appropriate safety glasses must be worn at all times** – the use of contact lenses is also discouraged, however, if you do plan to wear contact lenses in the laboratory you must inform your instructor and wear safety **goggles**.

2. **Proper dress is required** – Bare feet, sandals, or opened-toed shoes are not allowed in the laboratory. It is best not to wear expensive clothing as stains and holes can result from misplaced chemicals.

3. **Food and drinks of any kind are not allowed in the laboratory** – Keep all objects, such as glassware or plastic tubing, out of your mouth while in the lab.

4. **Behaving appropriately** – When working in the laboratory, you must be aware of others around you. Students will be carrying chemicals to and from their workstation, so be careful when walking through the laboratory. **In other words no horseplay in the lab.**

5. **Cleaning up** – It is important that you clean your workstation upon completing your laboratory exercise. Make sure the gas is turned off to the Bunsen burners, remove any paper towels to the trash bin and clean up any spilled chemicals.

6. **Learn the location and operation of the safety equipment**- Your laboratory is equipped with safety showers, eyewashes and fire extinguishers. You should become familiar with the location of each of these items as well as the location of all exits. If the fire alarm goes off while you are in the laboratory

turn off all open flames, and follow the instructions of your laboratory instructor.

7. **Safety video-** Before the first experiment begins all students will watch a safety video (**during lecture**) outlining the topics discussed above as well as a description of proper handling of chemicals and chemical spills. After watching the film each student will fill out and sign the safety sheet found at the end of this chapter. If you choose to wear contact lenses you must also complete the contact lens form.

Laboratory Techniques

In this section you will learn the proper technique for using some common laboratory equipment.

At your station you will find a graduated cylinder, burette, Bunsen burner and a 50-mL beaker. Your instructor will demonstrate the proper technique for reading the graduated cylinder and the burette as well as how to properly use a Bunsen burner and the analytical balances. After your instructor demonstrates these techniques you must demonstrate your ability to use this equipment to your instructor by completing the exercise below.

Competency Exercise

1. Light your Bunsen burner. Adjust the amount of air that is mixed with the gas so that a small bright blue cone is in the center of the flame and no "yellow" flame is visible. Show the flame to your instructor. Afterwards, fully turn off the gas source so that the flame of your Bunsen burner is extinguished.

2. Fill the graduated cylinder about half full of tap water. Record the volume of water (to one decimal place) in the Data section and ask your instructor to verify your answer.

3. Fill the burette with water to approximately the zero mark. Drain a little of the water by rotating the stopcock so that no air bubbles are in the tip, making sure the water level is below the zero mark.

4. Read the volume off the burette and record your answer Data Table 1-1. In this measurement, you can record two decimal places because the burette is more exactly calibrated than the graduated cylinder.

5. Using the analytical balance record the weight of the empty 50-mL beaker. Make sure the balance reads zero and is recording in grams. If it is not reading zero be-

168

fore placing the beaker on the balance push the Tare button to reset the balance to zero.

6. Using the water in the burette, fill the 50-mL beaker about quarter full. Read the new volume off the burette and record the value, again to two decimal places.

7. Using the balance, weigh the 50-mL beaker and added water and record your results.

8. Calculate the volume and weight of water placed into the 50-ml beaker by finding the difference between the initial and final volume and weight measurements. Record your answers on the data sheet.

9. Given that the **density** of a substance is its mass divided by its volume, calculate the density of the water. Record your answer on the Data Table 1-1.

170

Name _____   **Data Table 1-1**

Date_____

	Data
1. Volume of Water in Graduated cylinder: (Round to the nearest decimal place)	_____mL
2. Initial Volume of Water in the burette: (Round to the nearest two decimal places)	_____mL
3. Volume of water in the burette after filling the 50-mL beaker	_____mL
4. Volume of water added to the 50-mL beaker.	_____mL
5. Weight of the empty 50-mL beaker.	_____g
6. Weight of the 50-mL beaker and water	_____g
7. Weight of water added to the 50-mL beaker	_____g
8. Density of water (step 7/ step 4)	_____g/mL

Name _____   **Data Table 1-2**

Date_____

Safety in the laboratory Exercise

The following exercise should be attempted after you have read the introduction to this experiment and watched the safety video.

1. What non-clothing item must be worn in the laboratory at all times?

2. If the fire alarm goes off what do you do?

3. What is the proper procedure in the event of an acid spill?

4. How do you handle broken glass?

Name _____   Data Table 1-3

Date_____

In the space provided below draw a diagram of your laboratory
noting all exits and safety equipment.

Experiment 2

Physical Properties of Substances

Objective

To find the density of several substances, and determine the identity of an unknown substance based on its measured physical property.

Equipment and Chemicals

100-mL graduated cylinder   Salt-water solution
150 and 250-mL beaker
Unknown liquids
Unknown metals
Burette

Safety Precautions

Wear approved eye protection
Avoid inhaling chemicals.

Principles

Chemists measure the physical properties of substances, properties that can be measured without changing the sample, in order to distinguish one substance from another and to determine how the substance might be useful for some practical application. Some of the more important physical properties are briefly defined below.

Density of a Solid (A):

Use a large test tube, available on the side shelf, to collect an unknown metal from your instructor. The metal will be one of those listed in the density table below. Weigh a 150 mL beaker to the nearest 0.1 gram. Add the metal and reweigh the beaker. Record both weights in the Data Section. Fill a 100 mL graduated cylinder about half full with tap water and read the volume accurately to 1.0 mL (record in Data Table 2-1). Add the metal and record the new volume level of the water. Pour out the water and return the metal to your instructor.

Table of Density in (g/cm^3)

Element	Fe	Zn	Cd	Sn	Pb
Density	7.85	7.10	8.65	7.28	11.4

Density of a Liquid (B):

To calculate the density of an unknown solution, we need to measure the mass of a measured volume of solution. A 50 mL burette will be used to measure the volume. Clean and rinse your burette and allow it to drain. Obtain about 150 mL of one of the unknown solutions. Rinse your burette twice with 5 to 10 mL of this solution and drain. Fill the burette to about the 0 mL mark and drain out some liquid so that no air bubbles are in the tip and the solution is now slightly below the 0 mL mark. Read the burette accurately to 0.05 mL and record the data. Now weigh a dry 150 mL beaker. Add around 30-35 mL of the liquid from the burette to the beaker. Read the burette again and reweigh the beaker, recording all data in Data Table 2-1. Refill your burette and repeat the process again, only you do not need to rinse the burette as second time.

Table of Density in (g/mL)

Solution	Water	25% Salt Water	90% Ethanol
Density	0.978 g/mL	1.25 g/mL	0.86 g/mL

Name _____   **Data Table 2-1**

Date_____

A. Density of a Solid

1. Weight of empty beaker   _____ g
2. Weight of beaker + metal   _____ g
3. Weight of metal   _____ g
4. Volume of water   _____ cm^3
5. Volume of water + metal   _____ cm^3
6. Volume of solid   _____ cm^3
7. Density of Solid [(3)/(6)]   _____ g/cm^3
8. Name of solid (see table)   _____

B. Density of a Liquid

1. Initial burette reading   _____ cm^3
2. Final burette reading   _____ cm^3
3. Volume of liquid   _____ cm^3
4. Weight of beaker   _____ g
5. Weight of beaker + liquid   _____ g
6. Weight of liquid   _____ g
7. Density of liquid [(6)/(3)]   _____ g/cm^3
8. Identity of liquid   _____

Name _____   **Data Table 2-2**

Date_____

Questions

1. What error, if any, would the calculated density of the solid have if the material had a hollow center?

Name _____

Data Table 2-3

Date_____

Pre-laboratory Exercise

1. Calculate the density of an object that has a dry mass of 12.7 g and displaces 1.57 mL of water when submerged. (Show all work)

2. What mass of solid lead would displace exactly 234.6 liters of water? (hint use the table of densities)

3. A cup of gold colored metal beads was measured to have a mass 425 grams. By water displacement, the volume of the beads was calculated to be 48.0 cm3. Cm3 are equal to mL. Given the following densities, identify the metal.
   Gold: 19.3 g/mL
   Copper: 8.86 g/mL
   Bronze: 9.87 g/mL

188

Weight of Copper in Copper Sulfate Pentahydrate

Experiment

3

Objective

To measure the amount of copper in a copper compound and calculate the experimental and theoretical percent of copper found in the compound.

Equipment and Chemicals

$CuSO_4 \cdot 5H_2O$	250 mL Beaker
1.0 M H_2SO_4	50 mL Beaker
Zinc (granulated)	

Safety Precautions

Wear approved eye protection. Be careful to have no flames present in the laboratory after the zinc is added to the reaction mixture. The H_2 gas produced in this step can react violently with O_2 in the air in the presence of a spark or flame.

Principles

The percent by weight of any part of a substance is the weight of that part divided by the weight of the whole substance, times 100%. In order to experimentally obtain the weight of any part of a substance it must be separated from the whole. In this experiment, the reaction shown below where zinc is exchanged for copper will be used to separate the copper from a measured weight of copper sulfate pentahydrate.

$$Zn(s) + CuSO_4(aq) \rightarrow Cu(s) + ZnSO_4(aq)$$

This reaction is a successful procedure because the $ZnSO_4$ is soluble and the copper solid can be separated from the other components by decantation. An excess of zinc metal is used in the experiment to completely convert all of the copper ions to the metal in a short time period. The excess zinc can be removed before weighing the copper because it reacts with dilute sulfuric acid as shown below.

$$Zn(s) + H_2SO_4(aq) \rightarrow ZnSO_4(aq) + H_2(g)$$

189

Copper does not react under similar conditions and can be collected and weighed after the zinc is removed.

The percent copper in the sample is determined by dividing the mass of copper collected in the experiment by the weighed amount of the copper compound. For example, in an experiment similar to the one above starting with $CuCl_2$, if the mass of the copper compound was measures as 2.32 g and the mass of the copper collected measured as 1.09 g, the percent of copper would be:

$$\text{Mass percent copper} = \frac{1.09g}{2.32g} \times 100\% = 47.0\,\%$$

In this case we know the formula of the starting copper compound and can calculate what the theoretical value should be. Using the molar masses of the elements listed in the periodic table we can add the molar mass of Cu which is 63.55 g/mol and that of $CuCl_2$ which is 134.45 g/mol.

$$\text{Theoretical mass percent copper} = \frac{63.55 \text{ g/mol}}{134.45 \text{ g/mol}} \times 100\% = 47.27\,\%$$

This value is slightly higher than the calculated value, indicating that in the actual experiment, not all of the copper was collected.

Procedure

1. Weigh the empty beaker first. Then place a one scoop ~(1.5 grams) sample of copper sulfate pentahydrate into the 250 mL beaker and weigh it again. Enter the weights in the data section. Subtraction of the mass of the empty 250 mL beaker from the second mass (250ml beaker + $CuSO_4 \cdot 5H_2O$) yields the mass of $CuSO_4 \cdot 5H_2O$ transferred to the 250 mL beaker. This method is known as *weighing by difference* and is generally the preferred method.

2. Add 50 mL of 1 M H_2SO_4 to the 250 mL beaker, then warm gently and stir the mixture until the sample is dissolved.

3. *Turn off the flame* (the gas evolved in the next step is hydrogen) - no open flames should be near your reaction at this point because the reaction of H_2 with O_2 to make H_2O is very violent and is set off by a spark or flame. Compare with the reference test tube with around 1.2 grams of zinc metal and dump the similar level of zinc into a new tube. Place the zinc metal in the solution of copper sulfate and cover the beaker with a watch glass. The reaction should be done under the hood in order to evacuate H_2 gas ASAP. Allow the reaction to proceed, removing the cover every few minutes to stir the solution. The reaction is complete when the solution is colorless and no more gas is evolved. *If gas evolution still occurs*

when the solution is colorless, add 2 mL of dilute hydrochloric acid and stir and heat gently until gas evolution ceases.

4. Allow the copper to settle to the bottom of the beaker and carefully decant off liquid. Add 25 mL of water to the beaker and stir vigorously for a few minutes. Again allow the solid to settle to the bottom of the beaker and carefully decant the liquid.

5. After carefully pouring off the water, the residual moisture may be evaporated by heating the damp copper in the beaker over a Bunsen burner for approximately 45-60 seconds. Remove the beaker from the Bunsen burner once the copper has been dried adequately *(Use caution and the appropriate safety equipment as the beaker may be very hot).* Weigh the copper and beaker and record the value in the data section. Discard the copper into a beaker on the side shelf labeled copper waste. Make sure all of the copper is removed, reweigh the beaker, and record the weight in the data section. Perform the calculations.

As in any experiment of this type, some of the final product (copper in this case) will be lost before weighing. Your actual yield might be lower than the yield predicted (theoretical yield) from the masses of the starting materials (the limiting reagent). Thus your percent copper as found in the experiment may be lower than the theoretical amount. It cannot be higher than the theoretical amount if you weighed pure dry copper at the end and your initial mass of $CuSO_4 \cdot 5H_2O$ was correct.

192

Name _____   **Data Table 3-1**

Date_____

1. Mass of 250 mL empty beaker   _____ g

2. Mass of 250 mL beaker + $CuSO_4 \cdot 5H_2O$ _____ g

3. Mass of $CuSO_4 \cdot 5H_2O$ (2-1)   _____ g

4. Mass of 250 mL beaker + copper   _____ g

5. Mass of empty 250 mL beaker (Ditto 1)   _____ g

6. Mass of copper (4-5)   _____ g

Calculations:

Mass % of Cu = _____

194

Name _____   **Data Table 3-2**

Date_____

Questions

1.   Why is it important to turn the flame off before adding the zinc to the reaction beaker?

2.   What could cause the calculated, or measured mass percent to be greater than the theoretical percent?

Name _____   **Data Table 3-2**

Date_____

Pre-laboratory Exercise

1.   Calculate the theoretical mass percent of barium (Ba) in barium hydroxide, $Ba(OH)_2$

2. If all of the barium was removed from a 25.0 gram sample of barium hydroxide and the mass of the barium was 16.6 grams. Calculate the mass percent of barium and compare this to the theoretical mass percent calculated in question 1. Explain any differences

Experiment

4

Energy and Chemical Reactions

Objective

The objective of this experiment is to measure, using a basic calorimeter, the stored (potential) energy in an almond. You will use this data to calculate the nutritional energy, in Calories, of a bag of almonds.

Equipment and Chemicals

Unsalted Whole Almond
Large Tin Can
Water
Thermometer
Torch Lighter

Large Cork
Copper Wire
Glass Test Tube
Clamp

Safety Precautions

Be careful not to burn yourself when igniting the peanut

Principles

All chemical reactions are accompanied by a change in **energy**, which is defined as a property having the capacity to do work. The basic forms of energy include chemical, electrical, mechanical, nuclear and radiant or light energy. The two most important forms of energy are kinetic energy and potential energy. **Kinetic energy** is defined as the capacity to do work resulting from the motion of an object. For example, the movement of water through a dam results in the turning of a set of turbines. The movement of the water gives it the capacity to do work. **Potential energy**, which is defined as the energy stored in an object by virtue of its position. For example, a rock held one foot above ground has the potential to do work if it were dropped. However, that same rock would have the ability to do more work if it were elevated much higher

199

than a foot from the ground. Let's say it was dropped from 100 feet it would have the capability to do much more work. So, we could say that the rock held at 100 feet above ground would have a higher potential energy than the rock at 1 foot.

Just as the rock has stored energy due to its position, chemicals have stored energy due to its composition. For instance gasoline, which is a chemical, comprised of mostly carbon and hydrogen reacts with oxygen to form carbon dioxide and water:

$$Gasoline \quad + \quad O_2 \quad \rightarrow \quad CO_2 \quad + \quad H_2O + heat$$

As you know the energy released in this reaction is used to move the pistons in your automobile and make it move (producing work).

If the energy (usually in the form of heat) is released to the surroundings during a chemical reaction, as in the example of the combustion of gasoline, it is termed an **exothermic reaction**. On the other hand, some reactions need to absorb heat from their surroundings to proceed. These reactions are labeled **endothermic reactions**. A good example of an endothermic reaction is that which takes place inside of an instant 'cold pack'. Commercial cold packs usually consist of 2 compounds - urea and ammonium chloride in separate containers within a plastic bag. When the bag is bent and the inside containers are broken, the two compounds mix together and begin to react. Because the reaction is endothermic, it absorbs heat from the surrounding environment and the bag gets cold.

Reactions that proceed immediately when two substances are mixed together (such as the reaction of gasoline and oxygen) are called **spontaneous reactions**. However, not all reactions proceed spontaneously. For example, think of a match. When you strike a match you are causing a reaction between the chemicals in the match head and oxygen in the air. However, the match won't light spontaneously. You first need to input energy, which is called the activation energy of the reaction. This **activation energy** is defined, as the energy required to initiate a chemical reaction. So, in lighting the match you must first supply activation energy in the form of heat by striking the match; after the energy is absorbed and the reaction begins, the reaction continues until you either extinguish the flame or you run out of material to burn.

There are several units for energy, such as the **Btu** (British thermal units), which you probably recognize from air-conditioners and heaters. Two more commonly used units for energy are the **joule** and **calorie**. The calorie has been defined in many different ways. One accepted definition is the amount of heat needed to raise one gram of water by 1 degree Celsius. The calorie used in the context of food energy is actually kilo-calories (kcal) or a thousand calories and is often written as **Calories** with the c capitalized. So, for instance when you read the nutritional information on the back of a bag of potato chips and it reads 110 Calories it is actually 110,000 calories.

In this experiment you will construct a basic calorimeter- a device that directly measures the stored energy in a chemical. Using this device you will determine the nutritional Caloric content of a serving of peanuts. This is accomplished by heating a known mass of water and using a burning almond as the heating source. You will need to know the mass of the water heated. This can be determined from the fact that **water weighs 1.0 gram per milliliter**, and there are **236.6 ml in one cup**. So, one cup of water weighs 236.6 grams. Then knowing how many degrees the water temperature was raised, you can calculate how many calories are in the single almond. This will be an approximation because the entire almond will not be completely burned and there is still some chemical energy left inside the partially burned almond. However, this experiment will give a reasonable value of the stored energy. For example, let's say that the burning of one almond changed the temperature of 30 ml of water by 12.5 °C. This would require 375.0 calories of heat.

30 ml of water = 30 grams of water

(30.0 grams of water) x (12.5 °C) = 375.0 calories per almond

This of course would equal 0.375 nutritional Calories per almond.

Procedure

1. Using one end of a 4 inch piece of copper wire wrap the peanut and secure the long end into the cork.
2. Use the tin can as chimney around the almond on the copper wire. Position the almond so that it does not stick out of the top of the tin can chimney. above the bench with a wire gauze and ring stand
3. Using a ring stand attach a clamp to suspend the test tube directly above the almond. Leave approximately one quarter inch of space between the bottom of the

test tube and the almond. See Figure 4-1 for additional guidance in arranging your equipment.

4. Fill the suspended test tube with 15 mL of water.
5. Record the weight of the water by converting mL to grams (1 mL H_2O = 1 gram H_2O).
6. Put the thermometer into the water and record the temperature in Data Table 4-1. Leave the thermometer in the water.
7. Light the almond with the torch lighter while it is under the test tube of water and within the tin can chimney. The almond will take a few seconds to light make sure the entire almond is alight before ceasing the use of the torch lighter.
8. Allow the almond to burn out completely so that the water within the test tube can reach the maximum temperature possible.
9. Stir the water with the thermometer and record the temperature again.
10. Record the temperature in Data Table 4-1. Begin calculations for the experiment.

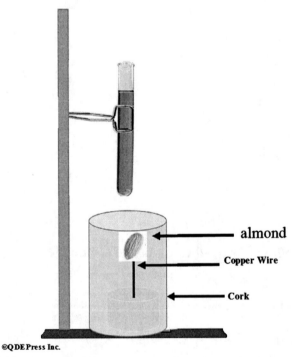

Figure 4-1

Name _____   **Data Table 4-1**

Date_____

1. Mass of the almond   _____grams

2. Mass of water   _____grams

3. Initial temperature of the water   _____°C

4. Final temperature of the water   _____°C

5. Temperature increase of water (Line 4-Line 3)   _____°C

6. calories in an almond (Line 5 x Line 1)   _____cal/gram

7. <u>C</u>alories in an almond (Line 6/1000)   _____Kcal/gram

8. Calculated Nutritional Calories per serving*   _____Cal/serving

9. Actual Nutritional Calories per serving   _____Cal/serving

*(look at the back of the package of almonds to determine serving size in grams)

204

Name _____   **Data Table 4-2**

Date_____

Questions

1. Discuss any potential flaws inherent with the equipment setup and general scheme of this experiment.

2. Why is it important to know the exact mass of water when conducting this experiment?

206

Name _____   **Data Table 4-3**

Date_____

Prelab

Define each of the following terms:

   1. Kinetic Energy

   2. Potential Energy

   3. Exothermic Reaction

   4. Endothermic Reaction

   5. Activation Energy

208

Experiment

5

Neutralization of an Acid with a Base

Objective

The objective of this experiment is to observe the base neutralization of an acid. This will be accomplished either directly (bubble formation) or indirectly (indicator color change).

Equipment and Chemicals

1 % Vinegar Solution
Graduated cylinder
Red cabbage indicator

Glass stirrer
Sodium Carbonate
500 mL beaker

Safety Precautions

Wear approved eye protection

Principles

An **Arrhenius acid** is a substance that is capable of releasing hydrogen ions (H^+) and producing hydronium ions (H_3O^+) in water solutions. In pure water there is a natural production of H_3O^+ and OH^- ions in equal numbers. However, when an acid is added to water, the number of hydronium ions increases. For instance when hydrochloric acid (HCl) is added to water the number of hydronium ions greatly increases:

$$HCl \quad + \quad H_2O \rightarrow \quad H_3O^+ \quad + \quad Cl^-$$
$$\text{acid} \qquad \qquad \text{water} \qquad \text{hydronium ion} \quad \text{chloride ion}$$

An **Arrhenius base** is defined as a substance that produces hydroxide ions, (OH^-) in water. And just as an acid can increase the number of hydronium ions a base will increase the number of hydroxide ions. For example when sodium hydroxide is added to water the number of hydroxide ions increase:

209

$$NaOH(aq) \rightarrow Na^+(aq) + OH^-(aq)$$

sodium hydroxide   sodium ion   hydroxide ion
(base)

Not only do acids and bases produce different ions in water, their properties differ as well. Acids generally taste sour, react with metals and all acids contain hydrogen. Whereas bases taste bitter, feel slippery and react with or breaks down fats and oils. There are many examples of acids and bases that we use in everyday life. For example vinegar (acetic acid), butter milk (lactic acid) oranges (citric acid) and of course stomach acid, this is hydrochloric acid. Bases found in the home include; ammonia (which is a solution of ammonium hydroxide), milk of magnesia (magnesium hydroxide) and your deodorants contain aluminum hydroxide.

The relative number of hydronium and hydroxide ions determines the strength of acids and bases respectively. Which simply means, the more of these respective ions in solution the stronger the acid or base. For instance the reason hydrochloric acid is a stronger acid than hydrofluoric acid is because adding the same number of molecules to water, HCl produces more hydronium ions than does HF.

The **pH** scale is a widely used method of describing the relative strengths of acids and bases. The term pH comes from the French *pouvoir hydrogene*, which translates to hydrogen power. pH is defined as the negative log of the hydronium ion concentration:

$$-log_{10}[H_3O^+]$$

In simple terms it is the logarithmic value of the number of hydronium ions in solution. It is not important, at this time that you understand the math involved in measuring the pH of an acid or base, but you should know that as the value of pH increases the strength of the acid decreases. Furthermore, the pH scale for all acids and bases spans from 0 to 14, where the strongest acid has a pH of 0.00 and the weakest acid is 14.00. Likewise, the strongest base has a pH value of 14.00 and the weakest base is 0.00.

A solution is said to be neutral- having the same number of hydronium ion as hydroxide ions when the pH is equal to 7.00. This of course is the pH of pure water. The following table gives a breakdown of different pH values:

pH	Definition
Above 7.00	Basic
Below 7.00	Acidic
7.00	Neutral
Near 0.00	Strongly acidic
Near 14.00	Strongly basic
Low pH	High hydronium ion (H_3O^+) amount
High pH	High hydroxide ion (OH^-) amount

pH indicators are used to determine the pH of a solution of acids and bases.

Acid-base indicators are generally large organic molecules that react with acids and bases. When they react, the structures of the molecules change and so does their color. Many colored solutions found around the home can be used as acid-base indicators: fruit and vegetable juices, food colorings, inks, and even tea. Some indicators such as litmus and phenolphthalein change color near the neutral point (pH = 7.00) as a solution changes from acidic to basic or vice versa. Others are one color in strongly acidic solution and a different color in weakly acidic and basic solutions. Still others pass through a variety of color changes as the pH changes. Red cabbage contains two principal types of plant dyes, anthocyanin and flavonol. The anthocyanin pigments are red in strongly acidic solution, blue in neutral and weakly basic solutions and colorless in strongly basic solutions. Weakly acidic solutions contain some of the red form and some of the blue form, and thus appear purple. Flavonol pigments are colorless in acidic and neutral solutions, and yellow in basic solutions. Weakly basic solutions thus contain both blue (anthocyanin) and yellow (flavonol) dyes and appear to be green. The color chart for red cabbage indicator is:

Red Cabbage Indicator Colors and pH Values													
Color	red		violet		purple		blue		green		yellow		
pH	1	2	3	4	5	6	7	8	9	10	11	12	13	14
-	Strong acidic		Moderate acid		Weak acid		Neutral		Weak base	Moderate base		Strong base		

However, red cabbage is not the only natural indicator that goes through a variety of color changes. Cherry juice, for example, may be red (pH = 2.5), orange (pH = 4.5), brown (pH = 7), or green (pH = 10). Litmus is a pigment obtained from lichens (a combination of algae and a fungus). It is red (pink) under acidic conditions and blue under basic conditions. Soaking paper in a litmus solution makes litmus paper.

Turmeric is a spice, which is made from ground turmeric tuber, and can be found alone or as an ingredient in curry powder. Turmeric is yellow in acid solutions and pink in basic solutions. Phenolphthalein is a chemical used by chemists as an acid-base indicator, and sold by pharmaceutical companies as a laxative. It is colorless in acid solutions and pink in basic solutions. Goldenrod paper contains a dye that is blue in concentrated solutions of strong acids, gold in other acidic solutions, and red in basic solutions.

In this experiment you will test a variety of household substances with an indicator you will prepare from red cabbage.

When an Arrhenius acid and Arrhenius base are combined, a chemical reaction occurs. This reaction is called a **neutralization reaction**. The products of the reaction are a **salt** and, usually **water**. The term neutralization comes from the fact that one product, water, is neutral, and the other, the salt, is often also neutral.

One such reaction occurs between hydrochloric acid (HCl, muriatic acid or pool acid) and sodium hydroxide (NaOH, lye). It can be written as follows:

$$HCl_{(aq)} + NaOH_{(aq)} \rightarrow NaCl_{(aq)} + H_2O_{(l)}$$

The **(aq)** means that the substance is dissolved in water to make an aqueous solution. An aqueous solution of table salt is neutral. If just enough sodium hydroxide solution is added to a sample of hydrochloric acid to react completely, the resulting solution is neutral. In practice, a small excess is added, and therefore, the resulting solution would be slightly basic.

In this experiment, the acid used will be acetic acid. Vinegar is a 5 % solution of acetic acid ($HC_2H_3O_2$) in water. Baking soda ($NaHCO_3$, sodium bicarbonate) will be used as the base. The salt produced is sodium acetate.

$$NaHCO_{3\,(aq)} + HC_2H_3O_{2\,(aq)} \rightarrow \square NaC_2H_3O_{2\,(aq)} + CO_{2\,(g)} + H_2O_{(l)}$$

Because the base used is bicarbonate, we will also generate carbon dioxide gas in the form of bubbles. The formation of these bubbles is a direct indication that an acid is still present. Once the acid has been neutralized the bubbling will stop. The baking soda may be used as an aqueous solution, as shown in the above equation, or as a solid. The results will be the same in either case.

Vinegar is acidic. As baking soda is added, the acid is gradually neutralized, until no more remains. The addition of more baking soda makes the solution basic. This neutralization will be monitored using a red cabbage indicator.

Procedure

Part I

1. Note the color and pH (from color chart) of the indicator and record it in the data table. This will be the color of a neutral solution.

2. Add 10.0 ml of vinegar and 15.0 ml of indicator and 10.0 ml of water to 500 ml beaker. Record the color of the solution in your data table.

3. Add 1 scoopula of baking soda solution to the above mixture, stir until all the baking soda has dissolved and record the color of the solution, and using the color chart record the pH of the mixture in your data table as well as the formation of any bubbles.

4. Adding 15 ml of baking soda at a time repeat step three 5 more times.

Part II

5. Predict the color of the indicator with only the baking soda and record your answer.

6. Take 10 ml of indicator and add to a 50 ml beaker.

7. Add 4 grams of baking soda to the indicator and record the color of the solution.

214

Name _____ **Data Table 5-1**

Date_____

Step 1. Color and pH of indicator solution:			
Steps 2-4. Scoops of baking soda added	**Bubbles** **(yes or no)**	**Color of the** **indicator**	**pH of the solution** **from the color chart**
0			
1			
2			
3			
4			
5			
6			
Step 5. Predicted color of the indicator with baking soda solution:			
Step 7. Experimental color of the indicator with baking soda solution:			

216

Name _____   **Data Table 5-2**

Date_____

Pre-laboratory Definitions

1.   **Arrhenius Acid**

2.   **Arrhenius Base**

3.   **Neutralization reaction**

Name _____   **Data Table 5-3**

Date_____

Review Questions

1. What is the color of indicator in acidic solutions?

2. What is the color of indicator in basic solutions?

3. Describe what happened to the vinegar solution as the baking soda solution was added.

4. Do you think that the change in color of the indicator and the lack of bubbles when more baking soda was added are related? Explain your answer.

Experiment

6

Wavelengths of Light Produced by Burning Ionic Compounds.

Objective

To determine the wavelengths of different colors of light produced by exciting electrons in different ionic compounds.

Equipment and Chemicals

5x50 ml beakers	copper chloride
Nichrome wire	copper sulfate
Bunsen burner	calcium chloride
Strontium chloride	sodium chloride
500 mL beaker	water

Safety Precautions

You will be using fire to light the chemicals. Be careful not to burn yourself!
Always wear safety goggles when in the lab. Avoid chemical contact with skin; some chemicals may cause irritation. Be careful when lighting the Bunsen burner and ask for instructor's help if needed. Always turn off Bunsen burner when finished. Keep face away from flame as some compounds may crackle when they come in contact with the flame.

Principles

Wavelength, represented by the Greek letter lambda (λ), is the distance between the peaks of two adjacent waves.

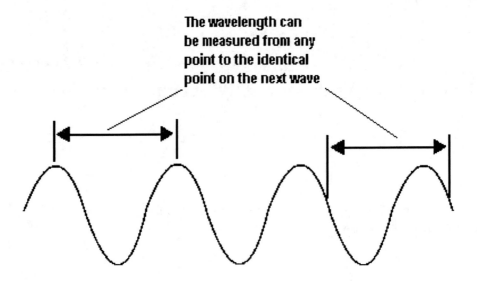

Wavelengths can be determined by using the equation, $\lambda = c/\nu$ where c is the speed of light (300,000,000 meters per second, or 3×10^8 m/s) and ν is the frequency in Hertz (Hz).

Because of this, the relationship between wavelength and frequency is inversely proportional, which is known as **dispersion relation**.

The **electromagnetic spectrum** is the range of all possible frequencies of electromagnetic radiation. Of these frequencies, the **visible spectrum** contains those visible to the naked eye and includes all colors of light except ultraviolet and infrared. Each color in the visible spectrum is due to a specific wavelength and frequency, denoted by the table below.

Color	Wavelength (λ) x10^{-9}	Frequency (ν) x10^{16}
Violet	4.1 meters	7.3 Hertz (Hz = 1/s)
Blue	4.8 m	?
Green	5.4 m	?
Yellow	5.8 m	?
Orange	6.1 m	?
Red	6.8 m	?

In this laboratory exercise, you will be asked to find the frequency of each color of light produced by burning several different ionic compounds.

For example burning potassium sulfate produces a violet flame which corresponds to a wavelength of 4.1 meters (table above). To find frequency, we rearrange the wavelength equation to:

$v=c/\lambda$, so $(3x10^8 m/s)/(4.1x10^{-9}m) = 7.3x10^{16} Hz$

The flame colors being produced by the ionic compounds are due to the input of thermal energy raising the electrons in the metal cation atom to a higher energy state. These electrons are referred to as being **excited**, and cannot remain in this state for very long. When the metal ion returns to their stable, or **ground state**, they will emit energy in the form of light.

Procedure

1. At your station should be five 50 ml beakers, each containing one of the ionic compounds and a fifth unknown.
2. Ignite your Bunsen burner with your instructor's help if needed. Then wet the end of your nichrome wire with water and dip it into one of the beakers, getting a little bit of the salt on the tip.
3. Then place the end of the wire containing the salt into the flame and observe and record the color.
4. At the end of each flame test, make sure to thoroughly rinse off the wire of any leftover compound before moving onto the next. Do this for each compound, recording each color observed. **Be sure to turn the Bunsen burner's gas source off when finished with the flame test!**

Name _____   **Data Table 6-1**

Date _____

Compound	Color	Wavelength x10^{-9}	Frequency x10^{16}
Strontium Chloride **SrCl$_2$**		_____ m	_____ Hz
Copper **Sulfate** **CuSO$_4$**		_____ m	_____ Hz
Calcium **Chloride** **CaCl$_2$**		_____ m	_____ Hz
Sodium **Chloride** **NaCl**		_____ m	_____ Hz
Unknown		_____ m	_____ Hz

Name _____   **Data Table 6-2**

Date_____

Questions

What might have caused a compound to produce a color different than expected?

Definitions

Define each of the following terms:

Wavelength

Dispersion relation

Electromagnetic spectrum

Visible spectrum

Ground state

Excited state

\

228

Name _____   **Data Table 6-3**

Date_____

Pre-laboratory Exercise

1. If a compound is burned and found to produce a flame with a wavelength of 5.15×10^{-9} m, what color would it resemble most?

2. A wave is known to have a frequency of 7.1×10^{16} Hz. What is its wavelength and what color does it resemble most?

    Wavelength (λ): _____ m
    Color:

230

Experiment

7

Shapes of Molecules

Objective

Use the VSEPR model to predict the shapes of molecules. To use these shapes to assign the hybridization of central atoms and the overall polarity of the molecules.

Equipment and Chemicals

Models of known and unknown compounds are provided on the side shelf.

Principles

The arrangement in space of the atoms in a molecule is an important component of its chemical bonding description. For example, CH_4 could be a flat planar molecule (incorrect) or tetrahedral (correct) as pictured below (the wedges mean that these H atoms are above the plane of the paper, the dash lines mean they are below with the carbon atom and hydrogen atoms connected by solid lines in the plane of the paper).

H
|
H—C—H
|
H

planar

H
|
C——H
H   H

tetrahedral

The chemical and physical properties of CH_4 and its many derivatives will be influenced greatly by the shape.

A simple model, based on Lewis structures, is extremely useful in predicting the shapes of many molecules; the valence-shell electron-pair repulsion (VSEPR) model. The main premise of the model is based on the idea that the electron pairs about an atom repel each other. The VSEPR model predicts the shape around each *central atom*, an atom in a molecule that is bonded to at least two other atoms. Starting with the Lewis structure, count the number of *lone pairs* on the central atom plus the number of *atoms bonded to it*. This sum determines the *electron-pair arrangement,* the shape that maximized the distances between the valence-shell electron pairs. For example, a central atom with two bonded atoms and no lone pairs will be linear because this shape puts the areas of electron density in the bonds as far apart as possible. Other shapes for electron-pair arrangements greater than two are in the textbook.

One important point needs to be emphasized before beginning the experiment. VSEPR theory predicts shapes based on the number of bonded atoms and lone pairs of electrons around a central atom; the *electron pair shape.* The actual *molecular shape* is frequently different because the lone pairs, although influencing the geometry, are not part of the final molecular shape. Molecular shape is always described in terms of the position of the atoms only. For example, the central nitrogen atom in NH3 has three bonded atoms and one lone pair; the electron pair shape is a tetrahedron. The molecular shape, as pictured below, is a trigonal pyramid because the area occupied by the lone pair is not part of the molecular shape. In contrast, BF3 has three bonded atoms and no lone pairs, so both its electron-pair shape and molecular shape are trigonal planar. The *electron pair arrangement* decides the actual *molecular geometry,* but is different from it if lone pairs are present on the central atom.

Procedure

Part I

Models of each of the compounds in this part are labeled on the side shelf. Draw a correct Lewis structure, count the number of atoms bonded to the central atom and the lone pairs on the central atom, and assign the correct electron pair shape to the first compound, NF3, listed in the Data Tables. Locate the model for that compound and draw a picture of it using wedges for a bond coming out or in front of the paper, a normal line for a bond in the plane of the paper and a dashed line for bonds projecting behind the paper. Use the model to name the molecular geometry (i.e., the ammonia pictured in the introduction was a trigonal (3-sided) pyramid). Carry out this procedure for all of the molecules or ions listed below.

Name _____   **Data Table 7-1**

Date_____

Molecule A NF$_3$	Lewis structure	Molecular shape
Number of atoms bonded to central atom		
Number of un-bonded electron pairs	Molecular shape drawing	
Electron pair shape		

Molecule B HCCl$_3$	Lewis structure	Molecular shape
Number of atoms bonded to central atom		
Number of un-bonded electron pairs	Molecular shape drawing	
Electron pair shape		

234

Name _____   **Data Table 7-2**

Date_____

Molecule C H_2CO	Lewis structure	Molecular shape
Number of atoms bonded to central atom		
Number of un-bonded electron pairs	Molecular shape drawing	
Electron pair shape		

Molecule D SiH_4	Lewis structure	Molecular shape
Number of atoms bonded to central atom		
Number of un-bonded electron pairs	Molecular shape drawing	
Electron pair shape		

236

Name _____   **Data Table 7-3**

Date_____

Part II

On the side shelf are six numbered models. Each one represents one of the molecules or ions in the below list. Assign a formula to each. Colors are not meant to represent any particular atom type.

H_2O	CH_4
SO_2	NCl_3
CS_2	HCN

Assign the molecules to the correct molecular shape.

1	2
3	4
5	6

Name _____   **Data Table 7-4**

Date_____

Pre-laboratory Exercise

1. Draw the Lewis structures for H_2O and HCN.

2. Fill in the table below.

Compound	Number of atoms bonded to central atom	Number of lone pairs on central atom	Electron pair shape	Molecular shape
H_2O				
HCN				

240

	Experiment
	8

Chemical Properties of Alkanes and Alkenes

Objective

In this experiment you will measure different chemical properties of different alkanes and alkenes and identify hydrocarbons based on their chemical properties.

Equipment

2- 10 cm test tubes
Aluminum foil
1% Bromine solution
1% potassium permanganate

Hexane
Hexane
evaporating dish
matches

Safety Precautions

Handle the 1% Bromine and the 1% potassium permanganate solutions with care. It can cause painful burns if it comes in contact with the skin.

Hydrocarbons are extremely flammable do not exceed the recommended amount of hydrocarbons you are to ignite. Be sure to get any hydrocarbon off your skin prior to lighting.

Principles

Alkanes and **alkenes** are classified as hydrocarbons, organic compounds containing hydrogen and carbon atoms only. These are further separated into two groups: **saturated hydrocarbons** (alkanes) - hydrocarbons with carbon-carbon single bonds only- and **un-**

241

saturated hydrocarbons (alkenes)-hydrocarbons with at least one carbon-carbon double bond. See Figure 8-1.

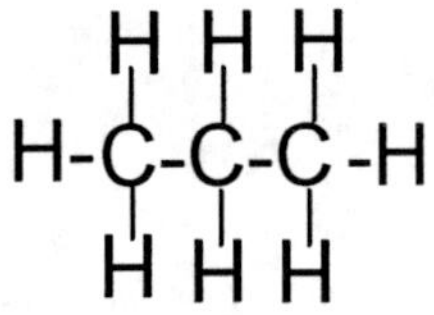

**Saturated Hydrocarbon
propane**

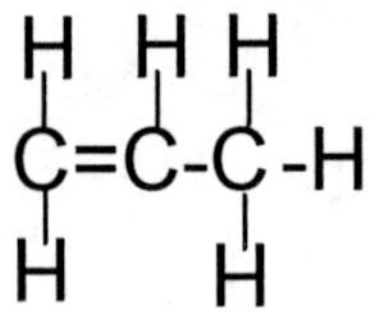

**Unsaturated Hydrocarbon
propene**

Figure 8-1

Hydrocarbons in different groups have different chemical properties. In general alkenes are more reactive than alkanes. For example both alkanes and alkenes react with bromine producing halogenated species. See Figure 8-2

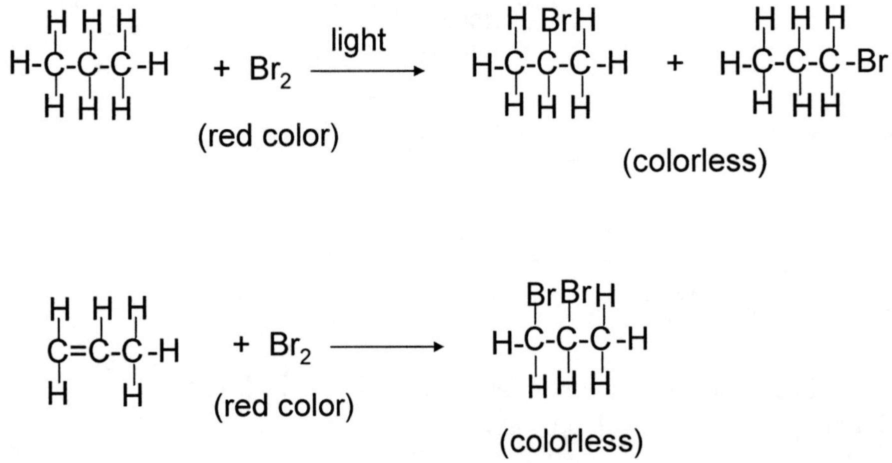

Figure 8-2

Notice from Figure 8-2 that in order for the alkane to react with bromine light is required, were as the alkene reacts easily with bromine. Another example of the difference in chemical reactivity is the oxidation with potassium permanganate. Alkenes are easily oxidized were alkanes are not. This can be observed by noticing the color of the permanganate solution changing from purple – (potassium permanganate solution) to a dark brown or black solution. This change in color comes from the formation of solid manganese dioxide during the oxidation process. See figure 8-3.

242

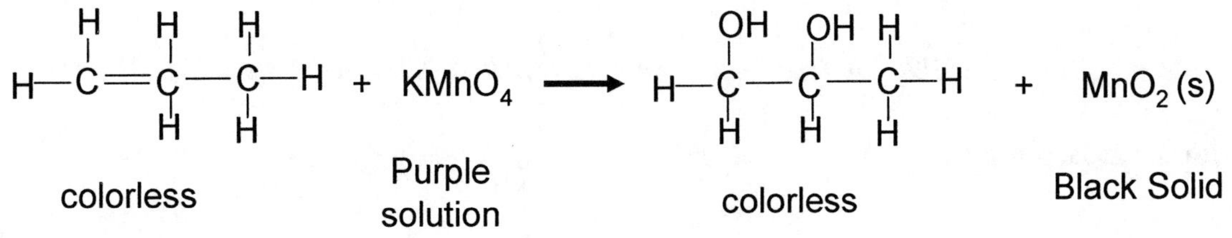

$$H-C=C-C-H \ + \ KMnO_4 \longrightarrow H-C-C-C-H \ + \ MnO_2 \,(s)$$

colorless   Purple solution   colorless   Black Solid

Figure 8-3

Combustion of hydrocarbons also shows different properties. Alkanes tend to burn cleanly whereas alkenes burn with the release of a dark smoke. Figure 8-4

$$Alkane + \ O_2 \ \rightarrow \ CO_2 + H_2O \qquad \text{Clean Blue-yellow flame}$$

$$Alkene + \ O_2 \ \rightarrow \ CO_2 + H_2O \qquad \text{Brown smoky flame}$$

Figure 8-4

You will now observe different chemical properties of both alkanes and alkenes by first reacting hexane and hexane with both bromine a solution of potassium permanganate and noting the differences. You will then do a simple flame test on both an alkane and an alkene and record your observations.

Procedure

(Remove all Data Tables for this experiment from the back of this book)

Reaction with Bromine

1. Obtain two test tubes and place 10-drops of hexane in one and 10-drops of 1-hexene in the other.
2. Wrap both test tubes with aluminum foil- to prevent light from entering the test tubes.
3. Add 2-3 drops of 1% Bromine solution to each test tube and stir with a stirring rod.
4. Wait about 1 minute and remove the aluminum foil. Record your observations in Data Table 8-1.
5. Observe the test tubes for a few more minutes and record any changes.

6. When finished pour waste into organic waste containers on the side shelf

Reaction with Potassium permanganate

1. Clean and dry test tubes from part 1.
2. Add 10-drops of methyl ethyl ketone (side shelf) to each tube.
3. Add 2-drops of hexane in one tube and 2-drops of hexene in the other tube
4. Add 2-frops of 1% potassium permanganate solution in each test tube and stir with stirring rod.
5. Record any changes you observe in Data Table 8-2.
6. When finished pour waste into organic waste containers on the side shelf

Hydrocarbon flame tests

1. In a clean dry evaporating dish place 5-drops of hexane
2. Carefully with a match ignite the hexane
3. Observe the color of the flame and or any smoke evolved.
4. record you observation in Data Table 8-2.
5. Repeat 1-4 with hexene

Identification of an unknown

1. Ask your TA for an unknown hydrocarbon. Use the experiments outlined above and determine if the unknown is an alkane or an alkene record your observations in Data Table 8-3. Follow all safety and disposal procedures as above.

Name _____   **Data Table 8-1**

Date _____

Reaction with Bromine

1.   Observation after removing foil   _____

       Hexane   _____

       _____

       _____

2.   Observation after removing foil   _____

       Hexene   _____

       _____

       _____

3.   Observation after a few minutes   _____

       Hexane   _____

       _____

       _____

4.   Observation after a few minutes   _____

       Hexene   _____

       _____

       _____

246

Name _____   **Data Table 8-2**

Date_____

Reaction with Potassium permanganate

1.   Observation after adding $KMnO_4$ _____

      Hexane   _____

                                             _____

                                             _____

2.   Observation after adding $KMnO_4$ _____

      Hexene   _____

                                             _____

                                             _____

Hydrocarbon flame tests

1.   Observation after burning   _____

      Hexane   _____

                                             _____

                                             _____

2.   Observation after burning   _____

      Hexene   _____

                                             _____

                                             _____

248

Name _____   **Data Table 8-3**

Date_____

Unknown Hydrocarbon

1.   Bromine reaction observation   _____

_____

_____

_____

2.   Observation after adding $KMnO_4$ _____

_____

_____

_____

3.   Flame test   _____

_____

_____

_____

4.   Identity of the hydrocarbon   Alkane   or   Alkene

249

250

Name _____   **Data Table 8-4**

Date_____

1. What is the difference between an alkane and an alkene?

2. Define saturated and unsaturated.

3. What chemicals are used in this experiment?

4. What safety precautions are advised

5. What is the method of disposal

252

Chapter 1

Chemistry and the World Around You

Where is Chemistry?

- Chemistry is EVERYWHERE.
- From the air we breath
- The food we eat
- Colors we see
- Fuels we use in our automobiles
- Medicines we take
- And found in all living organisms like "US"

What is Chemistry?

- The study of matter and the changes it can undergo

- Matter- is anything that has mass and occupies space

Scientific method

- Observation
- Hypothesis
- Experiments – (prove or disprove)
- Publication –(further experiments)
- Confirmation
- Application

The Scientific Method

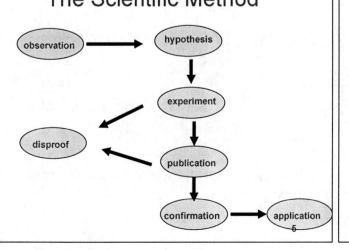

The Natural Sciences

- **Physical sciences-** geology, astronomy, physics and chemistry

- **Biological sciences-** Botany, zoology

Chemistry / Natural Sciences

- Geology – Geo-chemistry

- Astronomy – Cosmo-chemistry

- Physics – Chemical physics

- Chemistry – analytical, physical, organic, inorganic and biochemistry

Scientific research

- **Basic science-** research without the goal of a practical application

- **Applied science-** research with well defined, short term goals for a specific problem

- **Technology-** application of science for industrial production and societal goals

Chapter 2

The Chemistry of Matter

Key Questions

- What are the properties of matter
- How is matter classified
- What are elements and pure substances?
- What are compounds?
- What are mixtures?
- What are physical & chemical properties?

Key Questions Cont.

- What are physical & chemical changes?
- What are chemical reactions?
- How do we classify matter?
- How do we use the symbols of elements in chemical compounds and equations?
- What is quantitative vs. qualitative?

Types of matter

Macroscopic – large enough to been seen with the naked eye.

Microscopic- seen only with the aid of a microscope

Submicroscopic- can only be observed indirectly.

Properties of Matter

- **Mass-** is a measure of the amount or quantity of matter in an object

- **Weight** – is the force that results from the attraction between matter and the earth

Properties of Matter Cont.

- **Intensive property-** properties that are independent of the amount of the sample
  - Concentration, density, color, boiling point etc.

- **Extensive property** – depends on the size of the sample.
  - Mass, volume, energy etc.

Elements

Elements- a pure substance consisting of only one kind of atom (i.e. copper) and can not be decomposed into simpler substances by normal chemical means

Atoms- smallest unit of an element (i.e. copper)

Pure substance- matter with fixed composition at the submicroscopic level (i.e. copper)

Compounds

Chemical compounds– Pure substances composed of atoms of different elements combined in definite, fixed ratios. And can be decomposed into simpler substances or elements by chemical means

For example H_2O is a pure substance but not an element.

Compounds cont.

Once elements combine to form a chemical compound their individual characteristics are replaced by that of the compound.

Compounds cont.

For example in H_2O, a compound made from hydrogen and oxygen:

Hydrogen – flammable when burned in the presence of Oxygen

Water – used to put out fires

Mixtures

Most samples of matter occur in nature as mixtures which exists in two forms:

Homogeneous - uniform in composition
(i.e. milk)

Heterogeneous - not uniform in composition
(i.e. oil and water)

Solutions

Homogeneous mixtures of different compounds (which may or may not be pure substances) combined in such a way that the composition is uniform through out.

And can be in the solid, liquid, or gaseous state.

Purifying solution

Solutions can be purified by separation techniques.

For example a salt water solution can be changed into pure water by distillation (boiling and collecting the water vapor)

Separation

Separation of mixtures is accomplished by taking advantage of the different properties of the individual components.

For example water will boil at 100 °C where as salt will not.

Chemical and Physical Properties

Chemical property- properties that result in a chemical reaction converting the identity of one or more of the substances

Physical property – properties that can be observed or measured with out changing the identity of the substance

Chemical vs. Physical Properties

Chemical
– Flammability
– Anti-inflammatories
– biodegradable

Physical
– Boiling point
– Melting point
– Density
– Color

Chemical and Physical Changes

Chemical change- Process in which one or more pure substances are converted to one or more different pure substances.

Physical change- Process where the identity of the substance remains intact

Chemical vs. Physical Changes

Chemical
- Cooking beef
- Rusting
- Leaves changing color

Physical
- Boiling
- Melting
- Freezing

Classification of Matter

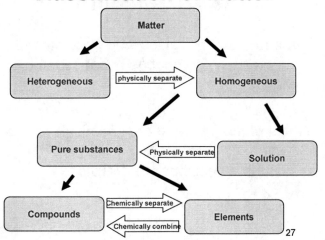

Chemical Reactions

Process in which one or more substances are converted to one or more different substances

Chemical Reactions Cont.

Reactants- Substances that undergo change in a chemical reaction.

Products- Substances formed as the result of a chemical reaction.

Hydrogen + Oxygen ⟶ Water
    Reactants   products

Molecules and Molecular Compounds

Molecule- The smallest unit of a chemical compound that can exist independently; composed of atoms held together by covalent bonds.

Molecular compounds- Compounds composed of molecules at the submicroscopic level.

Changes in Matter-energy

Energy-The capacity for doing work or causing change.

Potential energy- Energy in storage by virtue of position or arrangement.

Kinetic energy- The energy of objects in motion.

Quiz 2A

Which of the following questions would a chemist study?

a) Does an orange contain vitamin C ?

b) What gives a leaves their color ?

c) How many leaves does the average tree produce in a season?

Quiz 2A (cont.)

If a sample of matter contains only one kind of atom it is a(an) _____and_____broken down into simpler substances.

The two kinds of pure substances are _____ and _____

A solution is a(an) _____ mixture.

Quiz 2A (cont.)

If you melt ice in a glass, you have carried out a(an) _____ change because the ice changed into _____ .

Which of the following are chemicals?   a) Aspirin   b) Baking soda
c) Rubbing alcohol   d) Vanilla
e) Soap

Quiz 2A (cont.)

Which of the following mixtures are homogenous?
a) Sugar and water   b) Oil and water
c) Cement, sand, and gravel

Solutions may exist in the solid, liquid, or gaseous states.
a) True   b) False

Quiz 2A (cont.)

Identify by name the reactant(s) shown in the following equations:

$$CH_4 + 2O_2 \rightarrow CO_2 + 2H_2O$$

methane   oxygen   Carbon dioxide   water

Quiz 2A (cont.)

Which of the following are physical properties and which are chemical properties?
a) Flammability   b) Density   c) Souring Milk

Chemical potential energy is converted into light and heat in combustion.
a) True b) False

Symbols of Elements

Some symbols are single letters from the first letter in the elements name.

Others are two letters from the first two letters of the name with the first capitalized and the second lower case.

Eleven symbols are not derived from the modern name of the elements

Symbols of Elements Cont'd

- H- hydrogen
- C- carbon
- N- nitrogen
- O- oxygen
- F- fluorine
- Mg – magnesium
- Al – aluminum
- Ni – nickel
- Co – cobalt
- Ba - barium

Elements not derived from their modern name

- Sb-   Antimony
- Cu-   copper
- Au-   gold
- Fe-   iron
- Pb-   lead
- Hg-   mercury
- K-   potassium
- Ag-   silver
- Na-   sodium
- Sn-   tin
- W-   tungsten

Diatomic molecules

There are seven nonmetallic elements that exist as two-atom entities, these are called diatomic molecules.

- H_2- hydrogen
- O_2- oxygen
- N_2- nitrogen
- Cl_2- chlorine
- F_2- fluorine
- Br_2- bromine
- I_2- iodine

Chemical Formulas

Chemical formula- Written combination of element symbols that represents the atoms combined in a chemical compound.

Subscripts- In chemical formulas, numbers written below the line to show numbers or ratios of atoms in a compound.

Ex. in H_2O there are 2 atoms of hydrogen and 1 atom of oxygen

Common Molecular Compounds

Compound	Formula
Water	H_2O
Carbon monoxide	CO
Carbon dioxide	CO_2
Sulfur dioxide	SO_2
Methane	CH_4

Types of Chemical Formulas

Molecular formulas

Chemical formulas that represent molecules with atomic symbols and subscripts.

Structural formulas

Chemical formulas that show the connections between atoms in molecules as straight lines.

Structural formulas

H−O−H   Water, H_2O

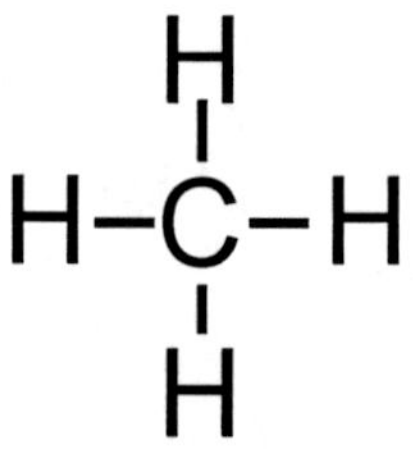

Methane CH_4

Chemical Equations

Chemical equations

Representations of chemical reactions by the formulas of reactants and products.

Balanced chemical equations

Chemical equations in which the total number of atoms of each kind is the same in reactants and products.

Chemical Equations cont'd

Coefficients- In a chemical equation are the numbers written before formulas to balance the equation.

Ex.

$2\ C(s) + O_2(g) \rightarrow 2\ CO(g)$

If no number is present assume 1 as the coefficient

Quiz 2B

Name the elements combined in the compound K_2HPO_4. How many atoms in total are represented in this chemical formula?

Quiz 2B (cont.)

Are the following equations balanced?

a) $2K(s) + Cl_2(g) \rightarrow KCl(s)$

b) $2Mg(s) + O_2(g) \rightarrow 2MgO(s)$

c) $SO_3(g) + H_2O(l) \rightarrow 2H_2SO_4(aq)$

Quantitative vs. Qualitative

Quantitative- Describes information or experiments that are numerical.

Qualitative- Describes information or experiments that are not numerical.

Base Units in the International System

Quantity	Unit	Abbreviation
Length	meter	m
Mass	kilogram	kg
Time	second	s
Temperature	kelvin	K
Amount	mole	mol

Prefixes Used With SI Units

Prefix	Abbreviation	Meaning
mega-	M	10^6
kilo-	k	10^3
centi-	c	10^{-2}
milli-	m	10^{-3}
micro-	μ	10^{-6}
nano-	n	10^{-9}
pico-	p	10^{-12}

Unit Conversion Factors

Unit conversion factor: a fraction in which the numerator is a quantity equal or equivalent to the quantity in the denominator, but expressed in different units

- 1 kg = 1000 g
  unit conversion factors: $\dfrac{1000\ g}{1\ kg}$ and $\dfrac{1\ kg}{1000\ g}$

Using Unit Conversions

- Express a volume of 1.250 L in mL, cm^3, and m^3

- $1.250\ L \times \dfrac{1000\ mL}{1\ L} = 1{,}250\ mL$

- $1.250\ L \times \dfrac{1000\ cm^3}{1\ L} = 1{,}250\ cm^3$

- $1{,}250\ cm^3 \times \dfrac{1\ m^3}{10^6\ cm^3} = 1.250 \times 10^{-3}\ m^3$

Quiz 2C

The prefix meaning 1000 times bigger, is _____, and the prefix meaning 0.001, or 1000 times smaller, is _____ .

Quiz 2C (cont.)

Which of the following experiments is qualitative and which is quantitative?

a) Determination of the distance between two atoms in a molecule.

b) Determination of the identity of the metal in a piece of wire.

Quiz 2C (cont.)

What are the units for the answer to the following calculations?

0.500ft x $\frac{12\text{ in.}}{1\text{ft.}}$ x $\frac{1m}{39.4\text{ in.}}$ x $\frac{1000mm}{1m}$ = 152___

Quiz 2C (cont.)

Which conversion factor should you use to convert miles to kilometers (km)?

a) $\frac{1\text{ mile}}{1.61km}$ b) $\frac{1.61\text{ km}}{1\text{ mile}}$

Accuracy and Precision

- **Accuracy** – is the term used to express the agreement of the measured value with the true value of the same quantity.

- **Precision**- expresses the agreement among repeated measurements

Chapter 3

A closer look at the Periodic Table and Atoms

Key Questions

- When, where and who developed the atomic theory?

- What is the structure of the atom?

- What are atomic and mass numbers?

- What are isotopes?

- What are atomic weights?

Key Questions cont'd

- Where are electrons located in the atom?

- How is the periodic table arranged?

- What are periodic trends?

- What are the main group elements and their properties?

John Dalton's Atomic Theory 1803

1. All matter is made up of small indivisible particles called atoms

2. Atoms of the same element are identical; Atoms of different elements have different properties

John Dalton's Atomic Theory Cont'd

3. Compounds are formed when atoms of different elements combine in small whole number ratios
4. Elements and compounds are composed of definite arrangements of atoms and a chemical change occurs when the atoms are rearranged

Dalton's theory was used to explain several scientific laws such as:

Law of Conservation of Matter

Matter is neither lost nor gained during a chemical reaction.

Antoine Lavoisier (1743-1794)

Dalton's theory also helped explain:

Law of Definite Proportions

In a compound, the constituent elements are always present in a definite proportion by weight.

Joseph Louis Proust (1754-1826)

Dalton's theory also helped explain:

- **Law of Constant composition-** all samples of a pure substance contain the same elements in the same proportions by mass.

- i.e. the chemical analysis of water finds a ratio of 8.0 grams of oxygen for every 1.0 grams of hydrogen

Structure of the Atom

There is now evidence of some 60 subatomic particles, however only three are needed to understand the make up of matter.

Electrons, Protons and Neutrons

Properties of Protons, Electron and Neutrons

	Relative charge	Relative mass	location
Electron	-1	0.00055	Outside the nucleus
Proton	+1	1.00727	Nucleus
Neutron	0	1.00867	Nucleus

Note: that the mass of a proton or neutron is about 2000 times that of an electron

The Nucleus of the Atom

- Experiments by **Ernest Rutherford (1871-1937)** determined that the atom is mostly empty space. However, over 99% of the mass is found in a very small region at the center of the atom called the **Nucleus**

- The diameter of the atom is some **100,000** times the diameter of the nucleus

Gold Foil Experiment

- Rutherford's experiment used beam of small subatomic particles (those smaller than the electron, proton or neutron) directed through a gold foil.

- Where he proved that most of these particles passed straight through without touching the foil

- This in turn lead to the modern view of the atom

Gold Foil Experiment

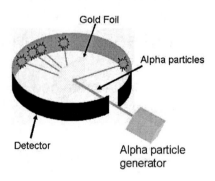

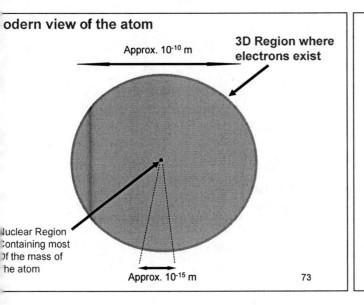

Modern view of the atom
- Approx. 10^{-10} m
- 3D Region where electrons exist
- Nuclear Region Containing most of the mass of the atom
- Approx. 10^{-15} m

Quiz 3A

- The law of conservation of matter states that matter is neither lost nor ___ in a(n) ___ reaction.

- The composition of sulfur dioxide (SO_2) is 32 parts by mass S and 32 parts by mass O. What is the percentage of sulfur in SO_2?

Quiz 3A (cont.)

- Hydrogen peroxide, H_2O_2, is always 94.12% O. What percentage of H in hydrogen peroxide?

- According to Dalton's atomic theory, a compound has a definite percentage by mass of each element because a) all atoms of a given element weigh ___, and b) all molecules of a given compound contain a definite number and kind of ___ .

Quiz 3A (cont.)

- What is the charge on the proton?
  a) –1   b) +1   C) 3

- Ernest Rutherford proposed the modern nuclear model of the atom.
  a) True   b) False

Quiz 3A (cont.)

- If the law of definite proportions is true, will the percent mass of silver, Ag, in silver sulfide be the same for all lumps or pieces of Ag_2S?
  a) Yes   b) No
- Natural gas is essentially methane (CH_4). Will methane produced from Texas gas fields have the same composition of the methane produced by gas fields in China? Explain your answer.
  a) yes   b) no

Quiz 3A (cont.)

- Most of the mass of an atom is concentrated in its
  a) nucleus   b) electrons   c) protons

- The mass of the proton is _____ times the mass of the electron.

Atomic Number & Mass Number

Atomic Number:
The number of protons in the nuclei of an atom.

Mass number → 19
Atomic number → 9 F ← Symbol of element
optional

Mass Number: Number of neutrons plus number of protons in the nuclei of atoms of an isotope.

Definitions

- **Atomic number (Z)** is the number of protons in the nucleus of an atom.
- **Mass number (A)** is the sum of the numbers of protons and neutrons in the nucleus.
- The **atomic number** determines the identity of the element; all H atoms contain 1 proton, all He atoms contain 2 protons, etc.

Isotopes

are different atoms of the same element that contain different numbers of neutrons (same Z, different A).

Hydrogen
1 proton
0 neutrons

Deuterium
1 proton
1 neutrons

Tritium
1 proton
2 neutrons

Symbols of Isotopes

- A symbol to identify a specific isotope is

$$^{A}_{Z}X$$

where A = mass number, Z = atomic number, and X is the one or two letter symbol of the element

- The three isotope of hydrogen are

$^{1}_{1}H \quad ^{2}_{1}H \quad ^{3}_{1}H$

Symbols of Isotopes

- Oxygen also has three isotopes, containing 8, 9, and 10 neutrons respectively. The symbols are

$^{16}_{8}O \quad ^{17}_{8}O \quad ^{18}_{8}O$

- Since the value of Z, and the symbol, both identify the element, Z is often omitted from the symbol

$^{16}O \quad ^{17}O \quad ^{18}O$

Example: Symbols of Atoms

- Write the symbol for the isotopes with:

- (a) 15 protons and 16 neutrons

- (b) 21 protons and 24 neutrons

Ions

- In many chemical reactions, atoms gain or lose electrons, producing charged particles called **ions**
  - A **cation** has a positive charge and forms when an atom *loses* one or more electrons
  - An **anion** has a negative charge and forms when an atom *gains* one or more electrons

Symbols for Ions

- The number of protons in the nucleus determines the symbol used for an ion
- The element's symbol is followed by a superscript number and a sign that shows the charge on the ion in electron charge units.
  - If the ionic charge is one unit, the number is omitted, e.g. Na^+ is the symbol for a sodium ion.

Example: Symbols of Ions

- Write the symbol for the ions that contain:

- (a) 9 protons, 10 neutrons, 10 electrons

- (b) 19 protons, 20 neutrons, 18 electrons

Example: Components of Ions

Fill in the blanks. $^{23}_{11}Na^+$
Symbol

Atomic number   ____
Mass number   ____
Charge   ____
no. of protons   ____
no. of neutrons   ____
No. of electrons   ____

Quiz 3B

- If a neutral atom has an atomic number of ten, then it has ____ protons and ____ electrons. If its mass number is 21, then it has ____ neutrons.

- In the symbol, $^{80}_{35}Br$ the number 35 is the ____, and the 80 is the _____ .

Quiz 3B (cont.)

- Isotopes of an element are atoms that have nuclei with the same number of ____ but different numbers of _____ .
- The negatively charged particles in an atom are ____ ; the positively charged particles are ____ ; the neutral particles are _____ .

Quiz 3B (cont.)

- In a neutral atom there are equal numbers of ____ and ____ .

- The number of protons per atom is called the ____ number of the element.

Quiz 3B (cont.)

- The diameter of the nucleus is ____ smaller than the diameter of the atom.

- An atom of arsenic, $^{75}_{33}As$, has ____ electrons, ____ protons, ____ neutrons.

Quiz 3B (cont.)

- The nuclei of all helium atoms contain exactly 2 protons.
  a) True   b) False

- All the isotopes of an element have the same relative abundance.
  a) True b) False

Atomic Masses and Atomic Weights

Atomic Mass Unit (amu)

The unit for relative atomic masses of the elements.

> 1 amu= 1/12 the mass of the carbon-12 isotope

Atomic Weight

The number that represents the (weighted average) atomic mass of the isotopes in a given element.

Isotopic Distribution

naturally occurring elements exists as a mixture of isotopes

i.e. percentage of hydrogen's three isotopes in a naturally occurring sample is 99.6% 1H, 0.25% 2H, and 0.15% 3H.

Atomic Weight Calculations

- An **atomic weight** is the weighted average mass in atomic mass units of an atom of an element. This is calculated by multiplying the fraction, that each isotope exists in a sample by their atomic mass units and adding each of the results together.

Atomic Weight Calculations

For example, the atomic weight of hydrogen would be:

(0.996 x 1) + (0.0025 x 2) + (0.0015 x 3)
= 1.0055µ or rounding gives **1.01µ**

This is also the number found under the symbol on the periodic table

Symbols on the Periodic Table

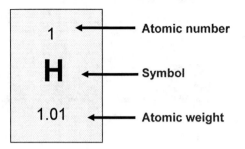

Electrons in Atoms

As stated earlier chemistry is the study of matter and the changes it undergoes

Many of these changes result in a chemical reaction involving the atoms of different samples of matter

Electrons in Atoms Cont'd

it is important to know that most, if not all, chemical reactions involve the electrons in one sample of matter interacting with the electrons in another sample of matter.

So, where are these electrons?

Location of Electrons in Atoms

- Interactions of light with different matter lead to important insight into the energy and location of electrons in atoms

Light has Wave Properties

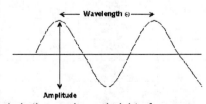

Amplitude is the maximum height of a wave

And Wavelength is the distance between
One peak and the next, measured in meters

Frequency- is the number of waves that pass a
Particular point in one second (hertz, Hz)

Continuous and Line Spectra

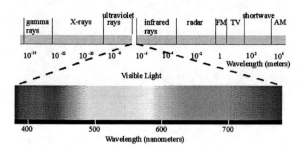

Line Emission spectra of Hydrogen

When high voltage is passed through a sample of gaseous hydrogen the electron in the atoms absorb energy (excited state). These excited atoms then emit light giving a unique spectra

Bohr Model of the Atom

In 1913 Niels Bohr assumed that electrons in atoms can exist only in certain energy states

Quantum

The smallest increment of energy that an atom can emit or absorb radiation.

Ground State

The condition of an atom in which all electrons are in their normal or lowest energy levels.

Excited State

An unstable, higher energy state of an atom

Ground state Energy levels in the Bohr atom

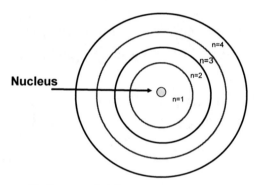

n= 1 is the ground state
n= 2 first excited state, etc.

Atom building – Bohr model

Level=n	$2n^2$	Max. # of e⁻
1	$2(1^2)$	2
2	$2(2^2)$	8
3	$2(3^2)$	18
4	$2(4^2)$	32
5	$2(5^2)$	50
6	$2(6^2)$	72
7	$2(7^2)$	98

n= principle shell

Bohr Arrangement of Electrons

- Phosphorous has 15 electrons in its neutral atom therefore it will have:
- n=1   = 2e   max =2
- n=2   = 8e   max =8
- n=3   = 5e   max =18

- So, Bohr arrangement is 2-8-5

Atomic Orbitals

Electrons occupying different energy shells (n levels) are located in atomic orbitals.

Where an **atomic orbital** is a region of 3-dimensional space where electrons exist around the nucleus

Subshells

Each principle energy shell consists of a certain number and type of orbitals grouped into **subshells**

The number of subshells within each principle energy level is equal in number to the value of n

Subshells Cont'd

For example, for the n = 1 principle shell there is only one subshell available and for the n = 3 principle shell there are three subshells available and are numbered 1, 2, and 3 respectively.

Subshells Cont'd

In other words the number of subshells available for each principle shell (*n*) is equal to 1, 2, 3, ...n

Atomic Orbitals

The number of atomic orbitals within each subshell is equal to 2 times the subshell number (Ss#) minus one:

Number of orbitals = (2 x Ss#) - 1

Atomic Orbitals Cont'd

For example, principle shell (n = 1) contains only one subshell and is numbered 1. Therefore the number of atomic orbitals in this subshell is:

$$(2 \times 1) - 1 = 1 \text{ orbital}$$

Atomic Orbitals Cont'd

The n = 2 principle shell has 2 subshells numbered 1 and 2. The number 1 subshell has only on atomic orbital, like in the previous example, however, for the number of atomic orbitals in the numbered 2 subshell is:

$$(2 \times 2) - 1 = 3 \text{ orbitals}$$

Atomic Orbitals Cont'd

Principle shell (n)	Subshells	Subshell letter	Orbitals per subshell
1	1	s	1
2	1, 2	s, p	1, 3
3	1, 2, 3	s, p, d	1, 3, 5
4	1, 2, 3, 4	s, p, d, f	1, 3, 5, 7

Electrons in Orbitals

The maximum number of electrons that can occupy any given atomic orbital is two

Subshell	# of atomic orbitals	Max # of electrons
s	1	2
p	3	6
d	5	10
f	7	14

Energy of Atomic Orbitals

The relative energy of atomic orbitals is based on the principle energy level (*n*).

However, as electrons are added around an atom the energy of the atomic orbitals tend to shift due to electron repulsion. This shift is exaggerated at the *n* = 3 principle energy level where the energy of the 3d subshell rises just above the energy of the 4s subshell.

Energy of Atomic Orbitals

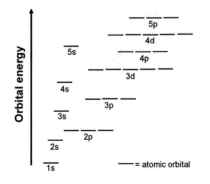

Electron Configuration

Electrons are added to atoms one at a time to the lowest available orbital. This process is called the **aufbau principle**

The distribution of these added electrons is termed the **electron configuration** for the atom

Electron Configuration

For example lets take an atom with six total electrons about the nucleus

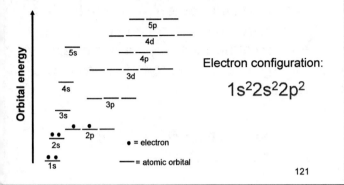

Electron configuration:
$1s^2 2s^2 2p^2$

• = electron
— = atomic orbital

Valence Electrons

Orbitals of the outermost or highest energy level and partially filled subshells of lower energy are called **valence orbitals**.

Electrons found in these valence orbitals are called **valence electrons**.

Example: Valence Electrons

How many valence electrons are in an atom with an electron configuration of
$1s^2 2s^2 2p^6 3s^2 3p^6 4s^2 3d^3$?

Example: Valence Electrons

How many valence electrons are in an atom with an electron configuration of
$1s^2 2s^2 2p^6 3s^2 3p^6 4s^2 3d^3$?

Highest occupied subshell = 2 electrons

Partially filled lower subshell = 3 electrons

Valence electrons = 5

Example: Valence Electrons

How many valence electrons are in an atom with an electron configuration of
$1s^2 2s^2 2p^4$?

Example: Valence Electrons

How many valence electrons are in an atom with an electron configuration of
$1s^2 2s^2 2p^4$?

Highest occupied orbitals = 6 electrons

There are no partially filled orbitals of lower energy level

Valence electrons = 6

Quiz 3C

- Which has less energy?
  a) Blue light   b) Red light
  c) Ultraviolet light   d) Infrared light

- Which color of light has the shortest wavelength?
  a) Blue   b) Green
  c) Orange   d) Red

Quiz 3C (cont.)

- According to Bohr's theory, light of characteristic wavelength is emitted as an electron drops from an energy level (closer to/farther from) the nucleus to an energy level (closer to/farther from) the nucleus.

- The maximum number of electrons in $n=3$ energy level is _____ .

Quiz 3C (cont.)

- The ground-state Bohr representation for electrons in an atom on K is _____ . K has _____ valence electrons.

- The ground-state Bohr representation of electrons in an atom of Cl is _____ . Cl has _____ valence electrons.

The Periodic Table

Dmitri Mendeleev (1834-1907) arranged the periodic table based on the Periodic law:

"When elements are arranged in the order of their atomic numbers, their chemical and physical properties show repeatable or periodic trends".

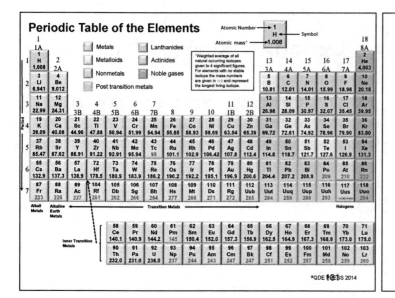

The Modern Periodic Table

Periodic Table: Arrangement of elements by atomic number in rows so that elements with similar properties fall together in vertical columns.

Groups: Vertical columns of elements in the periodic table with similar properties.

The Modern Periodic Table

Representative or main-group elements:
Elements in the A groups of the periodic table.

Transition elements: Elements in periodic table rows 4-7, or B groups in which *d* or *f* orbitals are being filled; they lie between main-group elements.

Inner Transition elements: Elements in which *f* orbitals are being filled, usually placed at the bottom of the periodic table (lanthanides and actinides).

The Modern Periodic Table

Periods– Horizontal rows in the periodic table.

Metals– Elements that conduct electric current; most are malleable and ductile.

Nonmetals– Elements that do not conduct electrical current

The Modern Periodic Table

Metalloids- Elements with properties intermediate between those of metals and nonmetals, and falling between them in the periodic table.

The Modern Periodic Table

Semiconductors
Metalloid elements with electrical conductivity intermediate between that of metals and nonmetals. Used in solid-state electronics

Noble gases
Group VIIIA elements in the periodic table.

Properties of Main Group Elements

- **Group 1A – alkali metals**
- **soft malleable metals, which can be easily cut with a knife**

- **react easily with all non-metals**

- **forms ions with +1 charge**

Properties of Main Group Elements cont'd

- **Group 2A- alkaline earth metals**
- **Harder than alkali metals, more dense and higher melting temperatures**
- **Very reactive but not as much as the alkali metals**
- **Form +2 ions**
- **All react with oxygen forming oxides**

Properties of Main Group Elements cont'd

- Group 7A – Halogens

- All exist as diatomic elements

- F_2 and Cl_2 are gases at room temp whereas Br_2 is a liquid and I_2 is a solid

---

Properties of Main Group Elements cont'd

- Group 8A- noble gases

- Single atom colorless gases at room temp

- Not very reactive, only very recently where these gases shown to react at all, this is why they are sometimes called **inert**.

Periodic Trends

Lets look at some trends that occur for across or down the periodic table

---

Atomic Properties Cont'd

- Atoms get **smaller** from left to right
- Atoms get **larger** from top to bottom

- **Reactivity**- for metals is related to size, the larger the atom the more reactive it becomes. For the nonmetals the opposite is true.

Atomic Radii

	1A	2A	3A	4A	5A	6A	7A	8A
							37	31
	Li	Be	B	C	N	O	F	Ne
	152	112	85	77	75	73	72	71
	Na	Mg	Al	Si	P	S	Cl	Ar
	186	160	143	118	110	103	100	98
	K	Ca	Ga	Ge	As	Se	Br	Kr
	227	197	135	122	120	119	114	112
	Rb	Sr	In	Sn	Sb	Te	I	Xe
	248	215	167	140	140	142	133	131
	Cs	Ba	Ti	Pb	Bi	Po	At	Rn
	265	222	170	146	150	168	(140)	(141)

Ionic Radii

Cations are smaller than their neutral atoms.

Anions are larger than their neutral atoms.

Ionic Radii Cont'd

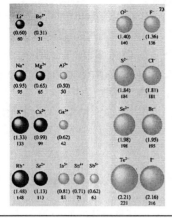

Ionization Energy

Is the energy needed to remove an electron from an atom. This energy is related to size of the atom, and increases from left to right, and decreases from top to bottom.

Ionization Energy Cont'd

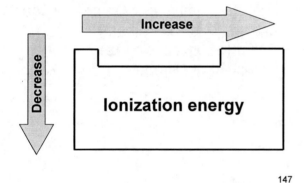

Quiz 3D

- How many valence electrons are in each of the following atoms?

a) Sodium(Na), Group 1A

b) Calcium(Ca), Group IIA

c) Boron(B), Group IIIA

d) Aluminum(Al), Group IIIA

Quiz 3D (cont.)

The modern periodic table has the elements placed in order of increasing atomic weight.

a) True   b) False

Quiz 3D (cont.)

- Metals typically have low numbers of valence electrons. a) True   b) False

- Nonmetals are typically found in Groups VA through VIIIA. a) True   b) False

Quiz 3D (cont.)

- Which of the following elements in Group IA is the most reactive: Li, Na, K, Rb, or Cs

- Group IA elements are all metals except for hydrogen. a) True b) False

- Isotopes have different numbers of valence electrons. a) True b) False

Quiz 3D (cont.)

- Which of the following are true about metals?
  a) They are found only in Group IA and IIA
  b) They are good conductors of heat.
  c) They are good insulators.
  d) They usually react to form positive ions.

Quiz 3D (cont.)

- Group IIA includes Be, Mg, Ca, Sr, Ba, and Ra. What is the predicted formula for the compound formed between Sr and Cl, if Be and Ba form $BeCl_2$ and $BaCl_2$?

Quiz 3D (cont.)

- Which of the following pairs of atoms has the greater number of valence electrons?
  a) Lithium(Li) or Oxygen (O)
  b) Sulfur (S) or Arsenic (As)
  c) Boron (B) or Nitrogen (N)

Quiz 3D (cont.)

- Which of the following pairs of atoms will be more reactive?
  a) Lithium (Li) or Cesium (Cs)
  b) Fluorine (F) or Bromine (Br)
  c) Beryllium (Be) or Calcium (Ca)

Chapter 4

Nuclear Chemistry, the chemistry of protons and neutrons

Key Questions?

- When was radioactivity discovered?
- What is a nuclear change?
- Why do some atoms undergo radioactive decay and others don't?
- What are the products of this decay?
- Why are some radioactive isotopes more harmful than others?
- What are some useful applications of this process?

Discovery

Henri Becquerel (1896) experimented with phosphorescence of certain minerals (uranium)

> It was later found that the radiation from these elements was alpha and beta particles and gamma rays.

Discovery Cont'd

Earnest Rutherford (1899)- found that alpha rays could be stopped by thin pieces of paper. Whereas beta rays were only stopped by at least 0.5 cm of lead.

Paul Villard (1900)- discovered the high energy, extremely penetrating gamma ray having characteristics of light waves. Very damaging to human tissue

Penetrating Abilities of Radioactive Particles

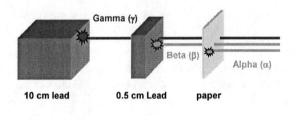

Madame (Marie) Curie (1859-1906)

Won the noble prize along with Henri Becquerel for their work on radioactivity.

She discovered that some elements are more radioactive than others.

Nuclear Reactions

Radioactivity- the result of a natural change of an isotope of one element into an isotope of a different element resulting in a nuclear reaction

Nuclear Reaction

Nucleons- protons and neutrons

During a nuclear reaction the number of nucleons is conserved but the identity of the element changes by emitting a particle or a ray.

Alpha Emitters

Radioactive decay of an atom resulting in the release of an Alpha particle and changing the identity of the atom

Alpha Particle- helium nuclei ($_2^4$He)

Alpha Emitters Cont'd

$$_{88}^{226}\text{Ra} \longrightarrow \ _2^4\text{He} \ + \ _{86}^{222}\text{Rn}$$

Radium- 226   Alpha particle   Radon- 222

Note- the mass number on the left equal the sum of the mass numbers on the right as well as the atomic numbers.

Beta Emitters

Radioactive decay of an atom resulting in the release of a Beta (β) particle and changing the identity of the atom

Beta (β) particle- an electron ($_{-1}^{0}$e)

Beta Emitters Cont'd

$$_{92}^{235}\text{U} \longrightarrow \ _{-1}^0\text{e} \ + \ _{93}^{235}\text{Np}$$

Uranium- 235   beta particle   Neptunium- 235

How does an electron come from the nucleus? An electron and a proton make a neutron.

$$_0^1\text{n} \longrightarrow \ _{-1}^0\text{e} \ + \ _1^1\text{H}$$

neutron   electron   proton

Gamma rays

After a nucleus emits an alpha or beta particle, it can still contain excess energy. This energy can be released in the form of highly energetic photons called **gamma rays (γ)**.

Gamma rays Cont'd

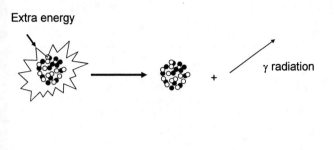

Positron Emission

A proton is converted to a neutron by positron emission

$$^{13}_{7}N \longrightarrow ^{0}_{+1}e + ^{13}_{6}C$$
positron

$$^{1}_{1}H \longrightarrow ^{0}_{+1}e + ^{1}_{0}n$$
proton   positron   neutron

Stability of Atomic Nuclei

Stability is based on relative number of protons and neutrons.

With the exception of Hydrogen isotopes, nuclei are stable when the mass number is at least twice as large as the atomic number.

Atomic Stability Cont'd

When a greater neutron/proton ratio exists (beta decay occurs)

When a greater proton/neutron ratio exists (positron emission occurs)

For elements greater than atomic number 83 (alpha emission occurs decreasing the number of protons and neutrons by 2)

Band of Nuclear Stability

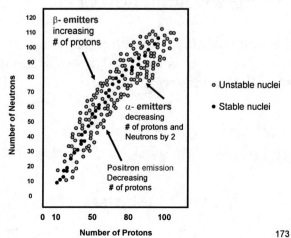

Summary of Radioactive Decay

Type of decay	Symbol	Charge	Mass	Change in atomic number	Change in mass number
Beta	$^{0}_{-1}e$	-1	0	+1	None
Positron	$^{0}_{+1}e$	+1	0	-1	None
Alpha	$^{4}_{2}He$	+2	4	-2	-4
Gamma	$^{0}_{0}\gamma$	0	0	None	none

Half Life

The rate of radioactive decay can be represented by its characteristic half life.

Half life- the time required for exactly 50% of the original material to decay.

Half Lives of Some Elements

Element	Half-life
Uranium-238	4.5×10^9 years
Hydrogen-3 (tritium)	12.3 years
Carbon-14	5730 years
Indium-131	8.05 days
Copper-64	12.9 hours
Zinc-69	55 minutes

Half Life Cont'd

Mass of radioactive isotope

60 g

30 g

15 g

7.5 g

1st half live → 2nd half life → 3rd half life →

Example Calculation of Half Life

Let's say you have a fresh sample of 100 grams of zinc-69 isotope which has a half life of 55 minutes. How much zinc remains after 165 minutes?

Answer

165 minutes / 55 minutes
  = 3 half lives = n
Therefore:

Fraction remaining $= \dfrac{1}{2^n} = \dfrac{1}{8}$

$\dfrac{1}{8}$ x (100 grams) = 12.5 grams of zinc remains

Example Calculation of Half life

For a particular isotope with a half life of 120 years, how long will it take for an 80 gram sample to decay to 10 grams?

80 grams → 10 grams = 3 half lives
Each half life = 120 years

3 x 120 = 360 years.

Quiz 4A

Which radioactive by product is most penetrating?

How is the nucleus of an isotope changed after beta decay?

The mass number of an isotope will decrease by 4 after which type of decay?

Quiz 4A Cont'd

What is a positron?

Who discovered gamma rays?

How long will it take for a 92 gram sample of copper-64 to decay to 23 grams, given the half life as 12.9 hours?

Applications of Radioactivity

Radio carbon dating- determining the age of a sample using the carbon-14 isotope

Gamma rays- from cobalt-60 and cesium-137 are used to irradiate food

Food radiation- retards the growth of organisms such as molds, bacteria, and yeasts.

.

Medicine

Radioactive isotopes are used in two distinct ways- diagnosis and therapy.

Diagnosis- radioisotopes are inserted into the patients body allowing an image to be produced of the problem area.

Medicine

Isotope	Name	Half-life hours	Uses
^{99}Tc	Technetium-99	6	Thyroid, brain, kidney
^{201}Tl	Thallium-201	21.5	heart
^{123}I	Iodine-123	13.2	thyroid
^{67}Ga	Gallium-67	78.3	Various tumors

Energy- Nuclear Reactions

Fission- large amounts of energy are released when heavy atomic nuclei split

Fusion- large amounts of energy are released when small atomic nuclei are combined.

Fission and Energy

About 25 million times more energy is released when a particular mass of uranium is used in a fission reaction compared to the same mass of natural gas during conventional combustion.

Nuclear Chain Reaction

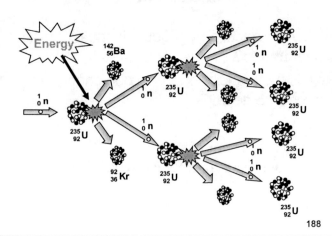

Electricity from Nuclear Reactions

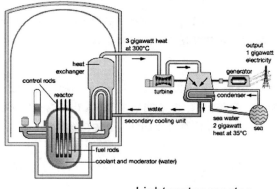

Light water reactor

Radioactive Waste

A major problem with fission energy is the amount of radioactive waste that must be stored. Some of which like plutonium is dangerously radioactive for thousands of years.

Radioactive Waste

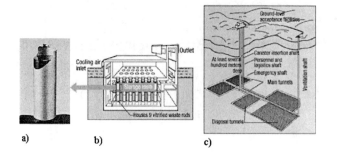

a) canisters for spent fuel rods
b) high level radioactive storage facility at the surface
c) underground radioactive waste facility.

Fusion and Energy

Fusion releases as mush energy as fission with fewer radioactive by-products

$$^{2}_{1}H + ^{3}_{1}H \longrightarrow ^{4}_{2}He + ^{1}_{0}n + 4.1 \times 10^{8} \text{ kcal of energy}$$

Approximately = 400,000,000,000 calories of heat

Quiz 4B

Name two uses of radioisotopes?

With the half-life of about 6 hours, how much Tc-99 would remain in a persons body after 18 hours if 12.0 grams was injected?

Quiz 4B Cont'd

Which nuclear energy process produces the least amount of radioactive waste?

Describe the differences in fusion and fission.

Why irradiate food stuff?

Chapter 5

Atomic Bonding

Key Questions?

- What are ionic bonds?
- What are the names and formulas for ionic compounds?
- What are polyatomic ions?
- What are covalent bonds?
- What are the names of covalent compounds?

Key Questions? Cont'd

- How do we predict the shape of molecules?
- What are polar vs. non-polar bonds?
- What are the properties on ionic and covalent compounds?
- What are intermolecular forces?
- What are the states of matter?
- What makes water unique?

Ionic Bonds

Octet Rule- In forming bonds, main-group elements gain, lose, or share electrons to achieve a stable electron configuration with eight valence electrons.

Ionic Bond- The attraction between positive and negative ions.

Ionic Bonds

Ionic Compounds: Compounds composed of positive and negative ions.

Formula unit: In ionic compounds, the simplest ratio of oppositely charged ions that gives an electrically neutral unit.

Sodium Chloride

"Table salt"

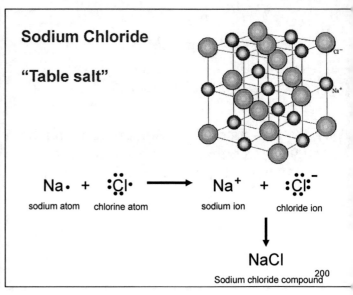

Predicting Formulas

Lewis dot symbols can be used along with the octet rule to predict formulas for ionic compounds.

Generally – metals in Groups IA, IIA, and IIIA react with nonmetals in Groups VA, VIA, and VIIA to form ionic compounds.

Common Ions

IA	IIA	IIIA	IV	VA	VIA	VIIA
H^+						
Li^+			$C^{\pm 4}$	N^{3-}	O^{-2}	F^-
Na^+	Mg^{2+}	Al^{3+}	$Si^{\pm 4}$	P^{3-}	S^{-2}	Cl^-
K^+	Ca^{2+}				Se^{-2}	Br^-
Rb^+	Sr^{2+}				Te^{-2}	I^-
Cs^+	Ba^{2+}					

Predicting Formulas for Ionic Compounds using Crossover Method

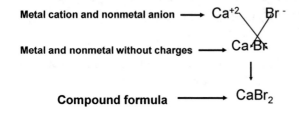

Naming Binary Ionic Compounds

Binary Compound- Chemical compound composed of one metal and one non-metal.

Cation = metal = common name

Anion = non-metal = name ends in –ide

Example - NaCl sodium chloride

Polyatomic Ions

- Polyatomic ion- a group of atoms with a net charge that behaves as a single particle

- The ammonium ion (NH_4^+) is the most common polyatomic cation

- There are many important polyatomic anions

Some Polyatomic Anions

Name	Formula	Name	Formula
Acetate	$CH_3CO_2^-$	Nitrate	NO_3^-
Carbonate	CO_3^{2-}	Nitrite	NO_2^-
Bicarbonate	HCO_3^-	Permanganate	MnO_4^-
Chlorate	ClO_3^-	Phosphate	PO_4^{3-}
Perchlorate	ClO_4^-	Hydrogen phosphate	HPO_4^{2-}
Chromate	CrO_4^{2-}	Dihydrogen phosphate	$H_2PO_4^-$
Cyanide	CN^-	Sulfate	SO_4^{2-}
Dichromate	$Cr_2O_7^{2-}$	Bisulfate	HSO_4^-
Hydroxide	OH^-	Sulfite	SO_3^{2-}

Example: Formulas Containing Polyatomic Ions

- Write the formulas of the compounds that contain:

- (a) the calcium ion and nitrate ion

- (b) the ammonium ion and the dichromate ion

Example: Formulas Containing Polyatomic Ions

- Calcium – Ca^{2+} Nitrate – NO_3^-

the combination has to result in a neutral species:

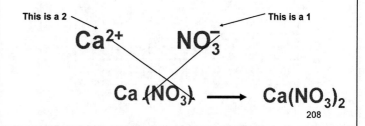

Quiz 5A

- The attractive forces between positive and negative ions in a crystal lattice are called ____ bonds.

- Positive ions are formed from neutral atoms by a) losing electrons, b) gaining electrons

- What charge is expected when the following atoms form ions?
  a) Lithium (Li)   b) Aluminum (Al)
  c) Sulfur (S)   b) Bromine (Br)

Quiz 5A (cont.)

- Which of the following atoms form positive ions?
  a) Potassium (K)   b) Bromine (Br)
  c) Nitrogen (N)   d) Sodium (Na)

- Negative ions are formed from neutral atoms by a) losing electrons b) gaining electrons

Covalent Bonds

- **Covalent Bond-** A bond in which 2 atoms share electrons to achieve a noble gas configuration.

- **Lewis Structure-** Electron dot representation of valence electrons in a molecule.

- **Bonding pair-** Pair of electrons shared between 2 atoms in a molecule.

- **Nonbonding pair-** Unshared pair of valence electrons in a molecule.

211

Lewis Dot Symbols

The valence electrons, represented by dots, are placed around the symbol until they are used up or until all 4 sides are occupied.

212

Lewis Dot Symbols

IA	IIA	IIIA	IVA	VA	VIA	VIIA	VIIIA
·H							He:
·Li	·Be·	·B·	·Ċ·	·N̈·	·Ö·	:F̈·	:N̈e:
·Na	·Mg·	·Al·	·Ṡi·	·P̈·	·S̈·	:C̈l·	:Är:
·K	·Ca·	·Ga·	·Ġe·	·Äs·	·S̈e·	:B̈r·	:K̈r:
·Rb	·Sr·	·In·	·Ṡn·	·S̈b·	·T̈e·	:Ï·	:Ẍe:
·Cs	·Ba·	·Tl·	·Pb·	·Bi·	·Po·	:Ät·	:R̈n:
·Fr	·Ra·						

213

Single Covalent Bonds

- Hydrogen atoms share their single electron giving them an electron configuration like helium (noble gas)

·H + ·H ⟶ H:H ⟶ H—H

214

Single Covalent Bonds

- Fluorine and hydrogen will share an electron each to obtain noble gas configuration:

·H + :F̈· ⟶ H—F̈:

215

Single Bonds in Hydrocarbons

Hydrocarbons
Compounds containing only carbon and hydrogen.

Alkanes
Hydrocarbons with carbon-carbon single bonds.

Saturated Hydrocarbons
Hydrocarbons that are alkanes.

216

Hydrocarbon single bonds

- One carbon atom and 4 hydrogen atoms will share electrons to achieve noble gas configuration:

$$4 \cdot H + \cdot \overset{\cdot\cdot}{C} \cdot \longrightarrow H-\underset{H}{\overset{H}{\underset{|}{\overset{|}{C}}}}-H$$

Multiple Covalent Bonds

Double Bond =
A bond in which 2 pairs of electrons are shared between atoms.

Triple Bond =
A bond in which three pairs of electrons are shared between atoms.

Double bond

- Ethylene C_2H_4 is formed from two carbon atoms and 4 hydrogen atoms, for each to have noble gas configuration a double bond must form between the two carbon atoms

$$4 \cdot H + 2 \cdot \overset{\cdot\cdot}{C} \cdot \longrightarrow \underset{H}{\overset{H}{\diagdown}}C=C\underset{H}{\overset{H}{\diagup}}$$

Bond Energy

Amount of energy required to break one mole (6.02×10^{23}) bonds between a specified pair of atoms.

Bond type	C-C	C=C	N-N	N=N
Bond length (nm)	0.154	0.134	0.140	0.124
Bond energy Kcal/mol	83	146	40	100

Electronegativity

- Is the ability of an atom to attract electrons toward itself.

- Electronegativity increases from left to right on the periodic table and from top to bottom.

Electronegativity Trend

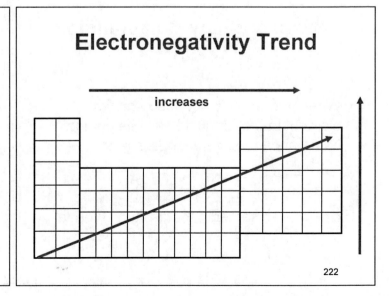

Electronegativity Cont'd

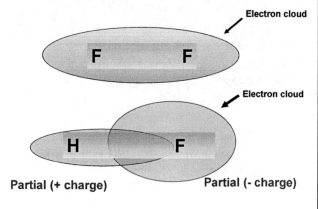

Polar and Non-polar Bonding

Non-polar
Describes a bond or molecule in which charge is evenly distributed, with no positive or negative regions. i.e. F_2

Polar
Describes a bond or molecule in which charge is unevenly distributed, creating positive and negative regions. Based on differences in electronegativity. i.e. HF

Strengths of polar bonds

- In general the further apart the atoms are form one another the greater the polarity of the bond.

- Ex. B-Cl is a more polar bond than N-Cl

- Compare their placement on the periodic table

Naming Binary Compounds

Hydrogen forms binary compounds with all non-metals (except noble gases).

Named as hydrogen followed by the non-metal name ending in -ide

Ex. hydrogen and fluorine is
$\qquad$ hydrogen fluoride

Naming Binary Compounds Cont'd

- When hydrogen is not used, the first element in the formula keeps its original name and the second ends in –ide

Ex. KCl is named   potassium chloride

Naming Binary Compounds Cont'd

- When more than one of the same atoms are used a prefix is added to denote the number of atoms in the name:

- Ex. N_2O is named dinitrogen monoxide

Naming Binary Compounds Cont'd

Prefixes for binary compounds

Number of atoms	prefix	Number of atoms	prefix
1	mono-	6	hexa-
2	di-	7	hepta-
3	tri-	8	octa-
4	tetra-	9	nona-
5	penta-	10	deca-

Naming Binary Compounds Cont'd

- If the first atom in the name is singular the prefix mono- is sometime omitted:

- Ex. CO is named carbon monoxide

Quiz 5B

- The name of SO_2 is _____

- The formula of sulfur trioxide is _____

- The name of HBr is _____

- The formula of dichlorine monoxide is _____

Quiz 5B cont'd

- The formula of sulfur dichloride is _____

- The name of $SiCl_4$ is _____

- How many electrons are shared in
a) double-bond b) triple bonds

- Which is the strongest bond?
a) C—C   b) C=C

Shapes of Molecules

- Shapes are determined from the number of bonding pairs and the number of lone pairs on the central atom.

- Use Lewis structures to determine shape

Shapes of Molecules Cont'd

Number of BP with no LP	Geometry	Ex.	Shape
2	linear	CO_2	O=C=O
3	trigonal planar	CO_3	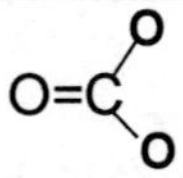 O=C(O)(O)
4	tetrahedral	CH_4	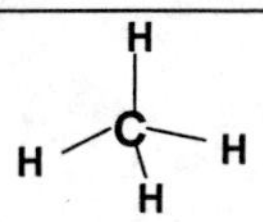 H—C(H)(H)—H

Shapes of Molecules Cont'd

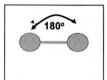 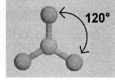 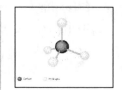

Properties of Molecular and Ionic Compounds

Most ionic compounds are electrolytes- (once dissolved in water the ions separate into positive and negative charged ions)

$$NaCl \xrightarrow{H_2O} Na^+_{(aq)} + Cl^-_{(aq)}$$

Comparison of Properties of Ionic and Covalent Compounds

- Crystalline solids (made of ions)
- High melting and boiling points
- Conduct electricity when melted or dissolved in water
- Most are soluble in water but not in nonpolar liquids

- Gases, liquids, or solids (made of molecules)
- Low melting and boiling points
- Poor electrical conductors in all phases
- Many soluble in nonpolar liquids but not in water

Properties of Molecular and Ionic Compounds Cont'd

Electrolyte– A compound that conducts electricity when melted or dissolved in water.

Non-electrolyte– A compound that does not conduct electricity when melted or dissolved in water, or does not separate into ions in water.

Electrolytes

If two wires connected to a light bulb is placed in pure water, the light will not come on.

Pure water

Electrolytes Cont'd

However, if sodium chloride is added to the water the light will turn on.

NaCl

Salt water solution

Quiz 5C

- a) An example of a molecule with covalent bonds in which the electrons are equally shared between the atoms is ____ .

- b) One where electrons are unequally shared is ___ .

Quiz 5C (cont)

- What is the geometry of CH_4?

- Which is a more polar molecule?
    a) H—H or H—F   b) C—H or C—O
    c) H—N or C—N

Intermolecular Forces

Intermolecular Forces- Attractive forces that act between molecules; weaker than covalent bonds

Dipole-Dipole forces- attractive forces between polar molecules

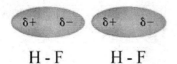

Intermolecular Forces Cont'd

Hydrogen Bonding- Attraction between a hydrogen atom bonded to a highly electronegative atom (O, N, F) and an electronegative atom in another or the same molecule.

H-X   X=O, N or F

Hydrogen Bond

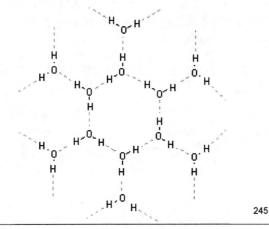

A comparison of boiling points for hydrogen and non-hydrogen bonded molecules

HB	BP °C	Non-HB	BP °C	Non-HB	BP °C
H_2O	100	H_2S	-50	H_2Se	-30
HF	20	HCl	-75	HBr	-55
NH_3	-10	PH_3	-80	AsH_3	-70

The States of Matter

Solids - Fixed shape and fixed volume, non-compressible, very strong intermolecular forces

Liquids - variable shape but fixed volume, strong intermolecular forces, non-compressible

Gases - Variable shape and volume, compressible, weak intermolecular forces

States of Matter

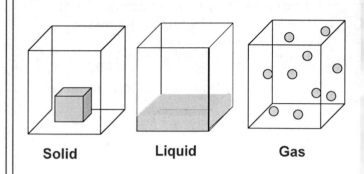

Solid Liquid Gas

Chapter 6

Chemical Reactions

Key Questions?

- What are balanced equations?
- What are moles?
- What are reaction rates and how do you influence them?
- What are equilibrium reactions?
- What are the first and second laws of thermodynamics?

Key Questions?

- What are oxidations and reductions?
- How do you recognize redox reactions?
- What are oxidation numbers?
- How do we use redox reactions?
- How do batteries work?

Balanced Equations

- A balanced chemical equation shows what happens during a chemical reaction

- **Remember**: the law of conservation of matter

$$CH_4(g) + 2O_2(g) \longrightarrow CO_2(g) + 2H_2O(l)$$

Balancing Equations Cont'd

$CH_4(g) + 2O_2(g) \longrightarrow CO_2(g) + 2H_2O(l)$

This tells us that one molecule of methane will react with two molecules of oxygen and will produce two molecules of liquid water and one molecule of carbon dioxide

Balancing Equations Cont'd

Write the correct formula for each substance
$$H_2 + Cl_2 \rightarrow HCl$$
Add *coefficients* so the number of atoms of each element are the *same* on both sides of the equation
$$H_2 + Cl_2 \rightarrow 2HCl$$

H 2 atoms   2 atoms
Cl 2 atoms   2 atoms

Balancing Equations Cont'd

Note

You can only add coefficients to balance chemical equations, you are never allowed to change the subscripts.

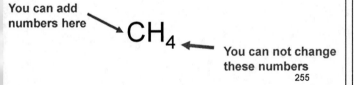

Balance the Following Chemical Equation

$$C_5H_{12} + O_2 \rightarrow CO_2 + H_2O$$

Assume one molecule of the most complicated substance, in this case C_5H_{12} and balance the others against it.

Balancing cont'd

- Adjust the coefficient of CO_2 to balance C

- $C_5H_{12} + O_2 \rightarrow 5CO_2 + H_2O$

- Then
- Adjust the coefficient of H_2O to balance H

- $C_5H_{12} + O_2 \rightarrow 5CO_2 + 6H2O$

Balancing cont'd

Finally :

Adjust the coefficient of O_2 to balance O

$C_5H_{12} + 8O_2 \qquad 5CO_2 + 6H_2O$

Check the balance by counting the number of atoms of each element.

What does this tell us?

$$C_5H_{12} + 8O_2 \longrightarrow 5CO_2 + 6H_2O$$

It tells us a lot, for one it tells us that one molecule of pentane C_5H_{12} will produce five molecule of CO_2, assuming you have enough Oxygen

Also:

If you needed 12 water molecules you would have to burn 2 molecules of pentane in the presence of excess oxygen

---

Molecules are "very" small

- Did you know that one molecule of pentane would only weigh 7.2×10^{-29} grams

- Do you know how many atoms of pentane you would need to fill a 12 oz. soda can?

Whoa!!!

You would need this many

3,053,000,000,000,000,00 0,000,000

What do we do?

- Since we can't weigh one atom at a time, what do we do?

We use the *MOLE*

The Mighty Mole

- One **mole** is the amount of substance that contains as many entities as the number of atoms in exactly 12 grams of the ^{12}C isotope.
- **Avogadro's number** is the *experimentally determined* number of the number of ^{12}C atoms in 12 g, and is equal to 6.022×10^{23}

The Mole

- One mole of anything contains 6.022×10^{23} entities

  1 mol H = 6.022×10^{23} atoms of H
  1 mol H_2 = 6.022×10^{23} molecules of H_2
  1 mol CH_4 = 6.022×10^{23} molecules of CH_4
  1 mol $CaCl_2$ = 6.022×10^{23} formula unit of $CaCl_2$

The Mole Cont'd

1 mol CH_4 = 6.022 x 10^{23} molecules of CH_4

also:

1 mole of CH_4 contains 4 moles of H

Ex. How many moles of O are in 3 moles of N_2O_4?

How do we use the Mole?

We obviously can't weigh one molecule of a substance, but we can weigh one mole of the same substance.

One mole of carbon weighs 12.00 grams of carbon.

Therefore we can say

C_5H_{12} + $8O_2$ ⟶ $5CO_2$ + $6H_2O$

- One mole of pentane will produce six mole of water

Molar mass

6	1
C	**H**
12.01	1.008

Number of grams in one mole of C = molar mass of C

Number of grams in one mole of H = Molar mass of H

So: one mole of CH_4 weighs 16.0 grams
Or: the molecular weight of CH_4 = 16.0 g/mol

Moles to Grams

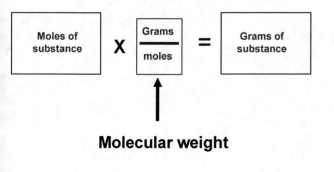

Molecular weight

Grams to Moles

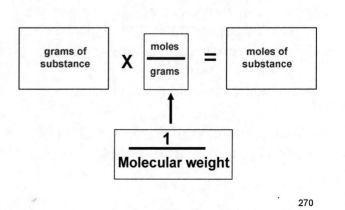

$\dfrac{1}{\text{Molecular weight}}$

Example Calculations

- How many grams are present in 0.35 mol of Na?
- How many moles are present in 34.5 grams of C_2H_6?

Example Calculations

How many moles of ethylene (C_2H_4, $\mathcal{M}$ = 28.0 g/mol) are present in 16 g ethylene?

Quiz 6A

- Balancing chemical equations is an application of the _____ .

- The _____ is used by chemists the same way the _____ is used by an egg farmer.

- Which is the molar mass of nitrogen (N_2) gas?
  a) 7g   b) 14g   c) 28g

Quiz 6A (cont.)

- In photosynthesis, carbon dioxide combines with water to form oxygen and the simple sugar glucose ($C_6H_{12}O_6$).

a) Balance the equation:

$$\_CO_2(g) + \_H_2O(l) \rightarrow \_C_6H_{12}O_6(aq) + \_O_2(g)$$

Quiz 6A (cont.)

b) How many molecules of CO_2 are needed to produce one molecule of glucose?

c) How many moles of CO_2 are needed to produce 1 mole of glucose?

d) What is the molar mass of glucose?

e) What is the mass in grams of CO_2 in needed to make 1 mole of glucose?

Rates of Reactions

Reaction rate- Amount of reactant converted to product in a specific amount of time.

What has to happen for a Reaction to Occur

1- Reactants have to come in contact

2- They must collide with enough energy to overcome the activation energy

3- they must have proper orientation when the collide

Activation Energy

Activation energy- Quantity of energy needed for successful collision of reactants; determines reaction rate

Activation Energy

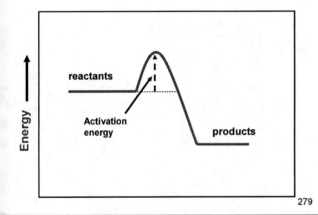

Controlling Reaction Rates

3 Strategies:
1. Adjust the *temperature*
2. Adjust the *concentration*
3. Add a *catalyst*

Catalyst: Substance that increases the rate of a chemical reaction by lowering the activation energy and without being consumed in the course of the reaction.

Catalyzed Reactions

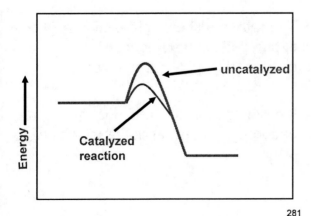

Chemical Equilibrium

Dynamic equilibrium- A state of balance between opposite changes occurring at the same rate.

Chemical equilibrium- Condition in which a chemical reaction and its reverse are occurring at equal rates.

Physical equilibrium – when the rate at which a substance changes between physical states (i.e. solid liquid and gas) is constant

Physical Equilibrium Cont'd

Ice bergs are constantly melting,
And at the same rate they are freezing

$$H_2O\ (s) \rightleftharpoons H_2O\ (l)$$

Chemical Equilibrium

Time ⟶

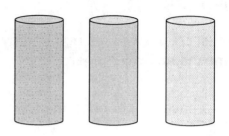

$$2NO_2 \rightleftharpoons N_2O_4$$

Le Chatelier's Principle

When a stress is applied to a system at equilibrium, the equilibrium shifts to relieve the stress.

Le Chatelier's Principle Cont'd

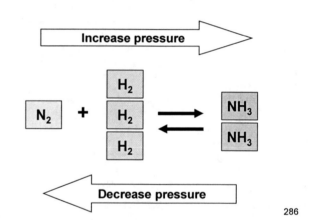

Le Chatelier's Principle Cont'd

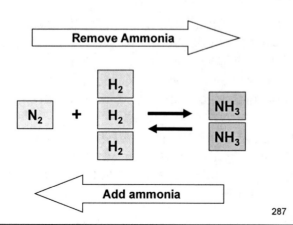

Quiz 6B

- The rate of a chemical reaction might be expressed as the amount of _____ converted to _____ per second.

- If colliding molecules do not have enough ___, they cannot react with each other.

Quiz 6B (cont.)

- A reaction can be speeded up by increasing either the ____ of a reactant or the ____ .
- A substance that speeds up a reaction without being a reactant or product is called a ____ .
- The ____ is like a hill that must be climbed to get a reaction going.

289

Quiz 6B (cont.)

- Are the amounts of reactants and products always equal to each other at equilibrium?
- Are the amounts of reactants and products present at equilibrium always the same for the same reaction under all conditions?
- At equilibrium, are the rates of the forward and reverse reactions equal?

290

What Drives a Reaction?

What causes some reactions to proceed spontaneously while others require a change in conditions in order to proceed?

Energy - heat in or heat out

Entropy- disorder or chaos

291

Driving Forces

Exothermic-
Describes a process that proceeds with the release of energy as heat. (results in a lower energy state)

Endothermic-
Describes a process that proceeds only with energy as heat from an external source. (results in a higher energy state)

Entropy-
A measure of the disorder of matter. If a change in entropy results in a more disordered state the entropy is said to be positive

292

Driving Forces Cont'd

- Exothermic and a positive entropy results in a spontaneous change

- Endothermic and negative entropy results in a non-spontaneous change. In other words has to be forced to react.

293

Laws of Thermodynamics

Thermodynamics- The science of energy as heat and its transformations.

First Law of Thermodynamics
Energy can be converted from one form to another but cannot be created or destroyed.

Second Law of Thermodynamics
The total entropy of the universe is constantly increasing.

294

Quiz 6C

- In an exothermic reaction, heat is _____

- In an endothermic reaction, heat is _____

- Some favorable chemical reactions start up as soon as the reactants are mixed, but others do not start till energy is added.
  a) True   b) False

Quiz 6C (cont.)

- When water freezes, its entropy _____ .

- Which of the following is a system with higher entropy?

  a) A new deck of cards as it comes from the box,

  b) A deck of cards after the cards are shuffled.

Quiz 6C (cont.)

- The two driving forces for favorable chemical change are a decrease in _____ and an increase in _____ .

- "You can't get something for nothing," is one way of stating the _____ .

Useful Reactions

The three types of reactions that are most beneficial to people of industrialized countries are:

Combustion
Oxidation-reduction
Neutralization

Combustion Reactions

A combustion reaction occurs when all substances in a compound combine with oxygen- commonly called burning

almost always exothermic

Combustion Reactions Cont'd

Gasoline, oil and wood, as well as many other common items that combust are called organic materials or compounds (Chapter 8).

made of carbon, hydrogen and oxygen

producing carbon dioxide, water and a lot of heat when burned

Combustion Reactions Cont'd

For example consider the combustion of methanol (Wood alcohol):

CH_3OH (l) + O_2 (g) → CO_2 (g) + H_2O (l) + Heat

Combustion Reactions Cont'd

However, many combustion reactions occur with compounds or elements that are not organic, like for instance the reaction of magnesium and oxygen:

$2Mg$ (s) + O_2 (g) → $2MgO$ (s) + Heat

*carbon dioxide and water are not produced

Applications of Combustion Reactions

fuel sources for cars, boats, planes, motorcycles, lawnmowers etc

combustion of natural gas and coal is used to heat and provide electricity for many homes and business

---

Oxidation Reduction Reactions

Oxidation- the gain of oxygen, the loss of hydrogen, or the loss of electrons

Reduction- the loss of oxygen, the gain of hydrogen, or the gain of electrons

- The species giving the electron is oxidized
- The species receiving the electron is reduced

Oxidation Reduction Reactions

When an electron is gained by a species it is said to be **reduced**

When a species loses an electron it is said to be **oxidized**

Redox Reactions Cont'd

The redox process occurs to a species when one of the following happens

Oxidation	Reduction
Loss of an electron	Addition of an electron
Addition of oxygen	Loss of oxygen
Loss of hydrogen	Addition of hydrogen

Redox Reactions Cont'd

As a simple example the reaction of elemental sodium with elemental chlorine:

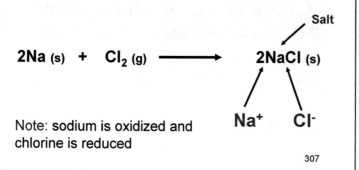

Note: sodium is oxidized and chlorine is reduced

Redox Reactions Cont'd

Combustion is another form of a redox reaction.

$$CH_4 \text{(g)} + O_2 \text{(g)} \longrightarrow CO_2 \text{(g)} + H_2O \text{(l)}$$

Note: **oxygen was added to carbon forming an oxide (oxidation)**

Oxides

Compounds of oxygen combined with another element. Most metals are mined as metal ores or oxides:

Metal	Formula	Name
Aluminum	Al_2O_3	bauxite
Iron	Fe_2O_3	hematite
Tin	SnO_2	Cassiterite

Recognizing Redox

- Redox reactions have to happen simultaneously, If there is oxidation there must be reduction they do not happen independently.

- To determine what is oxidized and what is reduced we have to determine oxidation numbers.

Oxidation Numbers

- Oxidation number are either positive or negative numbers assigned to each atom in a particular molecule. If the number changes from reactant to product then redox has occurred.

Assigning Oxidation numbers

1. Oxidation states for elements in uncombined form is zero

2. The oxidation state of a monatomic ion is the charge on the ion

3. In compounds F is always –1, the other halogens are –1 unless combined with a halogen above it, or oxygen, then it is +1.

Assigning Oxidation numbers Cont'd

4. In compounds H is +1.

5. In compounds O is always –2 unless combined with F then it is +2

6. The sum of the oxidation states must be zero for a neutral compound or equal to the charge on a polyatomic ion.

Example of Assigning Oxidation Numbers

- What is **oxidized** and what is **reduced**?

$$CH_4 (g) + O_2 (g) \longrightarrow CO_2 (g) + H_2O (l)$$

Example of Assigning Oxidation Numbers Cont'd

First separate each atom by drawing a line between them

$$C|H_4| (g) + 2|O_2| (g) \longrightarrow |CO_2| (g) + 2|H_2O| (l)$$

Example of Assigning Oxidation Numbers Cont'd

Then using the rules to give oxidation number to each element

$$\overset{-4\ +4}{C|H_4|} (g) + \overset{0}{|O_2|} (g) \longrightarrow \overset{+4\ -4}{|CO_2|} (g) + 2\overset{+2\ -2}{|H_2O|} (l)$$

Overall charge is zero

3 C must be -4
1 H is +1 there's 4 so total +4 Rule 4
elemental O is zero Rule 1
C must be -4
O is -2 1 There's 2 Of them so -4 Rule 5
Rule 4
Rule 5

One More

- Assign oxidation number to each atom in:

$$SO_4^{-2}$$

Just like before

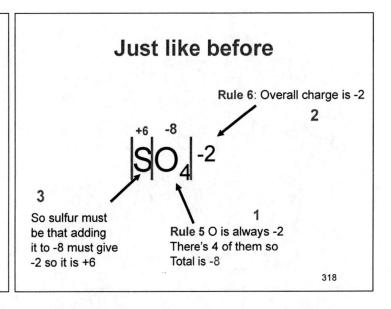

Rule 6: Overall charge is -2

+6 -8

$$|SO_4|^{-2}$$

3 So sulfur must be that adding it to -8 must give -2 so it is +6

1 Rule 5 O is always -2 There's 4 of them so Total is -8

Oxidizing and Reducing Agents

Oxidizing agent- A reactant that causes oxidation of another reactant by excepting an electron. and itself becomes reduced

Reducing agent- A reactant that causes reduction of another reactant giving it an electron and itself becomes oxidized.

Bleaching agent- A chemical that removes unwanted color by oxidation of the colored chemical.

Table of Oxidizing Agents

Name	Formula	Uses
Oxygen	O_2	Metabolism of foods, and combustion
Lead dioxide	PbO_2	Automobile batteries
Sodium Hypochlorite	NaOCl (aq)	Laundry Bleach disinfectant
Chlorine	Cl_2	Purification of water

Table of Reducing Agents

Name	Formula	Uses
Hydrogen	H_2	Fuel, chemical synthesis
Sulfur dioxide	SO_2	Chemical synthesis
Carbon	C	Iron Production
Zinc	Zn	Batteries

Oxidizing Agents: Free Radicals

Free Radical– A very reactive atom or molecule that contains an unpaired electron causing damage to anything it comes in contact with (like DNA) by removing an electron from it.

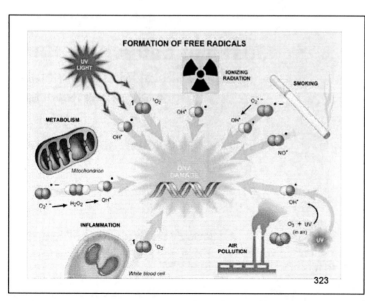

Anti-oxidants

Molecules that donate an electron to a free radical before it can cause damage:

Example of some common antioxidants:

Vitamins, E & C, beta-carotene (carrots) selenium (red onions)

artichokes, cranberries, blueberries, pecans and even cinnamon

Quiz 6D

Which is more oxidized? a) CO b) CO_2?

A chemical that causes reduction to take place is called a(n) _____ .

- The conversion of Na to Na^+ is an oxidation reaction. a) True b) False

Quiz 6D (cont.)

When nitrogen reacts with hydrogen to form ammonia, the nitrogen is said to be a) oxidized b) reduced

A chemical that causes oxidation to occur is a(n) _____ .

Quiz 6D (cont.)

- Which is the oxidized form of zinc, a) Zn^{2+} ion b) Zn atom

-

- In chemical definitions terms, an antioxidant is a(n) _____ .

- A disinfectant is a chemical classified as a(n) _____ .

Quiz 6D (cont.)

In the reaction $2Li(s) + S(s) \rightarrow Li_2S(s)$, sulfur is the oxidizing agent.
a) True b) False

In a chemical definition terms, an antioxidant is a(n) _____ .

The freeing of a metal from its ore usually requires a chemical reactant that is a(n) ___.

Quiz 6D (cont.)

The freeing of a metal from its ore usually requires a chemical reactant that is a(n) ___.

What is the oxidation states of each element in the following compounds?
H_2O, NF_3, PCl_5 MnO_4^-

Batteries

Electrochemical cell
A device in which a chemical reaction generates an electric current.

Battery
A series of electrochemical cells that produces an electric current; commonly refers to any electrochemical, current-producing device.

Batteries cont'd

Electrodes
Conducting materials at which electrons enter and leave an electrochemical cell.

Anode
Electrode at which electrons flow out of an electrochemical cell and oxidation takes place.

Cathode
Electrode at which electrons flow into an electrochemical cell and reduction takes place.

Electrochemical Equation

$$Zn(s) + Cu^{2+}(aq) \rightarrow Zn^{2+}(aq) + Cu(s)$$

Note that $Zn(s)$ is oxidized and Cu^{2+} is reduced

Oxidation of zinc and reduction of Cu

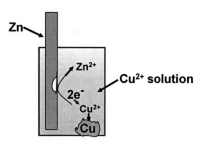

What if we separate the Zn and the Cu $^{2+}$ ions?

Electrochemical Cell

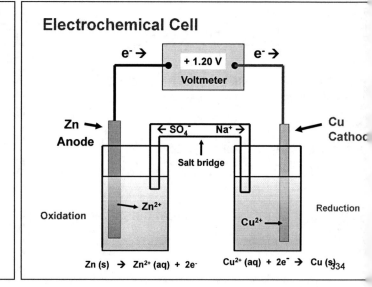

Types of Batteries

Primary battery
A battery that cannot be recharged because its reaction is not easily reversible; a throw-away battery

Secondary battery
A battery that can be recharged by reversing the flow of current, which reverses the current-producing reaction and regenerates the reactants.

Carbon-Zinc Cell (Primary Battery)

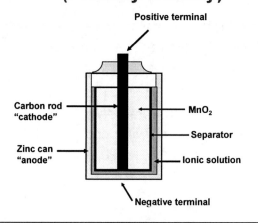

Lead Storage Battery

- nic solution
- ₂SO₄

Anode filled with lead

Cathode filled with PbO₂

Corrosion or Rusting

Corrosion- The unwanted oxidation of metals during exposure to the environment.

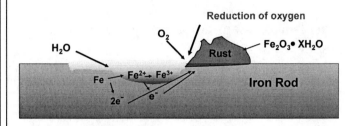

Oxidation of Iron

Quiz 6E

- A battery contains two electrodes, named the ____ and the ____ .

- It is necessary for ____ to flow between the electrode compartments in a battery so that ____ balance is maintained.

Quiz 6E (cont.)

The conversion of copper ions (Cu^{2+}) to metallic copper is ____ and will occur at the ____ in an electrochemical cell.

Name the three reactants needed for rusting iron.

Chapter 7

Acids and Bases

Key Questions?

- What are the properties and definitions of acids and bases?
- What are strong and weak acids and bases?
- What is a neutralization reaction?
- What is pH?
- What are buffers and their importance?
- What is acid rain? And how is it formed?

Acid-Base Definitions

Early definition of acids and bases was based upon their observable properties

- acids like vinegar or lemon juice, taste sour
- Bases on the other hand form solutions with a bitter taste and feel slippery to the touch.

Arrhenius Acid-Base Definition

Arrhenius acid is any substance that produces a hydrogen ion (H^+) when dissolved in water and the

Arrhenius base is any substance that produces the hydroxide ion (OH^-) when dissolved in water.

Arrhenius Acids & Bases

Acids- A substance once dissolved in water releases H^+ ions. (or H_3O^+)

Bases- A substance once dissolved in water releases OH^- ions.

$$HCl \xrightarrow{H_2O} H^+ (aq) + Cl^- (aq)$$

$$NaOH \xrightarrow{H_2O} OH^- (aq) + Na^+ (aq)$$

Bronsted-Lowry Definition

acid is a proton (H^+) donor and a

base is a proton acceptor

in order for a substance to act as an acid there must be a base present because if a substance is to donate a proton, there must be another substance present to accept it.

Bronsted-Lowry Definition

$$HCl_{(aq)} + NH_{3\ (aq)} \rightarrow NH_4^+{}_{(aq)} + H_3O^+{}_{(aq)}$$

Acid　　　**Base**

Properties of Acids and Bases

Acids	Bases
• Sour taste	• Bitter taste
• Give H^+ ions (Arrhenius)	• Provide OH^- (Arrhenius)
• Donate a proton (Bronsted-Lowry)	• Accept a proton (Bronsted-Lowry)
• React with metals to give Hydrogen	• Slippery feeling
Examples;	Examples;
vinegar, tomatoes, citrus fruit, and aspirin	ammonia, baking soda, soap, and detergents

Acid and Base Strength

- Strength of acids and bases are based on their ability to completely dissociate in water:
- Strong acids and bases completely dissociate in water, weak acids and bases establish an equilibrium when dissolved in water.

Acid and Base Strength Cont'd

- Strong acids:

$$HCl + H_2O \xrightarrow{100\%} H_3O^+ + Cl^-$$

- Strong base

$$NaOH + H_2O \xrightarrow{100\%} Na^+ + OH^- + H_2O$$

Weak acids and Bases

Weak acid:

$$HF + H_2O \underset{}{\overset{\sim 20\%}{\rightleftharpoons}} H_3O^+ \; F^-$$

Weak Bases

$$NH_3 + H_2O \underset{}{\overset{\sim 15\%}{\rightleftharpoons}} NH_4^+ + OH^-$$

Table of Strong Acids and Bases

Acid Name	Formula	Base Name	Formula
Hydrochloric acid	HCl	Sodium Hydroxide	NaOH
Nitric acid	HNO_3	Potassium Hydroxide	KOH
Hydrobromic acid	HBr	Lithium hydroxide	LiOH
Sulfuric acid	H2SO4	Magnesium hydroxide	$Mg(OH)_2$
Hydroiodic acid	HI	Calcium hydroxide	$Ca(OH)_2$

Acids and Bases

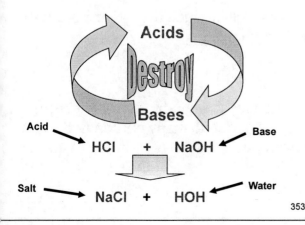

Acid → HCl + NaOH ← Base

Salt → NaCl + HOH ← Water

Neutralization Reactions

When a solution of a strong acid is mixed with a solution of a strong base the hydronium ion H_3O^+ in the acid reacts with the hydroxide ion OH^- in the base producing water two H_2O.

$$H_3O^+ + OH^- \longrightarrow 2H_2O$$

Neutralization Reactions Cont'd

- When a solution of an acid such as HCl and a solution of a base such as NaOH are mixed the result will be water and a salt.

HCl + NaOH ⟶ H₂O + NaCl
                              ↑
                            Salt

Neutralization Reactions Cont'd

- A salt is formed when the positively charged ion from the base combines with the negatively charged ion from the acid

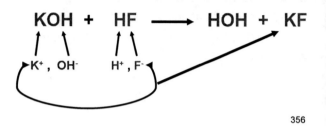

KOH + HF ⟶ HOH + KF
 K⁺, OH⁻   H⁺, F⁻

Neutralization Reactions Cont'd

- If the number of hydronium ions is equal to the number of hydroxide ions the solution will be completely neutralized.

- If there are more hydronium ions than hydroxide ions the solution will remain acidic

- If there are more hydroxide ions than hydronium ions the solution will remain basic

Acid-base Indicators

Substances that change color with changes in acidity or basicity of a solution.

Indicators Cont'd

○ Acid
■ Base

Increasing base amounts

Quiz 7A

- An acid will produce a ____ and a base will produce a ____ .

- To produce a desirable sour taste, a(n) ____ is added to many carbonated beverages.

Quiz 7A (cont.)

- Soap solutions are slippery and taste bitter, which indicates that soap is a(n) ___ (basic/acidic) substance.

- Complete the following equation for the reaction of an acid with a base.

$$H_3O^+(aq) + \underline{\hspace{1.5cm}} \rightarrow 2H_2O(l)$$

Quiz 7A (cont.)

- Identify each of the following chemical compounds as an acid, a base, or a salt:
  a) Na_2SO_4 b) HF(aq) c) KCl d) KOH

- Ammonia is a ___ base because it establishes an ___ with water in which ___ are formed.

Quiz 7A (cont.)

Complete the following equation:

$$\underline{\hspace{1.5cm}} + KOH\ (aq) \rightarrow KCl\ (aq) + \underline{\hspace{1.5cm}}$$

Concentration, Molarity

Concentration of a solution- The quantity of a solute dissolved in a specific quantity of a solvent.

Molarity [M]- Number of moles of solute per liter of solution.

M= $\dfrac{\text{moles of solute}}{\text{Liters of solution}}$

Calculating Molarity

- Calculating molarity requires knowing the number of moles of solute and the volume of the solution.
- Ex. If there are 2.5 moles of a solute in 3300 ml of solution the molarity would be:

$$M = \frac{\text{moles of solute}}{\text{liters of solution}} = \frac{\text{2.5 moles of solute}}{\text{3.3 L of solution}} = 0.76\ M$$

Calculating Molarity Cont'd

- Calculate the molarity of a solution made by dissolving 3.5 g (mw=40 g/mol) of NaOH in enough water to make 1.6 liters of solution.

Calculating Molarity Cont'd

First we must calculate the number of moles:

3.5 g NaOH × $\frac{1 \text{ mole NaOH}}{40 \text{ grams}}$ = 0.0875 moles NaOH

Then divide the number of moles by the liters of solution:

$\frac{0.0875 \text{ moles}}{1.6 \text{ liters}}$ = **0.054 M**

---

pH

pH is a numerical measure of the acidity or basicity of a solution. Or to be more specific it is a measure of the hydronium ion concentration.

By definition:

pH = Negative logarithm of the hydronium ion concentration of a solution.

---

pH Cont'd

In a **Neutral solution** the concentration of

the $[H_3O^+]$ = $[OH^-]$ = **1.0 ×10^{-7} M**

Taking the negative log of this number gives

pH = 7.00

By the way

- Bracketing a species indicates the "**concentration of**" that species

- For example $[H_3O^+]$ is read as "the concentration of hydronium ion"

Again remember: $[OH^-]$

"Brackets indicate concentration of"

---

pH Cont'd

Therefore if pH = 7.0, solution is *neutral* and the $[H_3O^+]$ = $[OH^-]$

if pH < 7.0, solution is *acidic* and $[H_3O^+]$ > $[OH^-]$

if pH > 7.0, solution is *basic* and $[H_3O^+]$ < $[OH^-]$

The pH of things

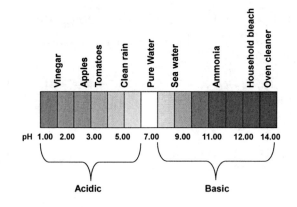

Relationship of pH and $[H_3O^+]$

The relationship between pH and [H3O+] is inversely and logarithmic.

as the pH decreases by one unit the hydronium ion concentration increases ten times. Conversely as the pH increases by one unit the hydronium ion concentration decreases ten times.

$[H_3O^+]$	pH	Decrease in $[H_3O^+]$ or Increase in $[OH^-]$
1.0×10^0	0	
1.0×10^{-1}	1	10 x
1.0×10^{-2}	2	100 x
1.0×10^{-3}	3	1000 x
1.0×10^{-4}	4	10000 x
1.0×10^{-5}	5	100000 x
1.0×10^{-6}	6	1000000 x
1.0×10^{-7}	7	10000000 x
1.0×10^{-8}	8	100000000 x
1.0×10^{-9}	9	1000000000 x
1.0×10^{-10}	10	10000000000 x
1.0×10^{-11}	11	100000000000 x
1.0×10^{-12}	12	1000000000000 x
1.0×10^{-13}	13	10000000000000 x
1.0×10^{-14}	14	100000000000000 x

Example, pH calculation

When a 12 oz can of Pepsi© is poured into a glass containing the same volume of water the pH of the water changes from 7.00 to 4.00. By what factor does the hydronium ion concentration increase in the water?

Example, pH calculation

Answer

The pH has decreased by 3 pH units from 7.00 to 4.00. Since each unit decrease in pH results in a 10 fold increase in $[H_3O^+]$ concentration. We can therefore say the increase in hydronium concentration is;

$10 \times 10 \times 10 = 10^3 = 1000$ times

Acid-Base Buffers

A buffer is– A solution of a weak acid and a weak base

A buffer does- maintains the pH of a solution

How does it do this?

• If an acid is added to a buffered solution, the base component of the buffer will destroy the added acid.

• If a base is added to a buffered solution the acid component of the buffer will destroy the added base

Components of Buffers

The acid and base have to be conjugate partners

Conjugate partners -the product formed, other than $[H_3O^+]$ or $[OH^-]$, from the dissociation of either a weak acid or weak base.

Conjugate Partners

when the weak acid HF is dissolved in water the products are the fluoride ion and the hydronium ion;

$$HF + H_2O \rightarrow F^- + H_3O^+$$

HF and F- are conjugate partners. Therefore, a solution of HF and F- results in a buffered solution.

Blood pH

The buffering components found in blood are carbonic acid (H_2CO_3) and the bicarbonate ion (HCO_3^-). In your blood, these weak acid-base partners will react with any added acid or base

Blood pH

When an acid is added:

$$H_3O^+ + HCO_3^- \rightarrow H_2CO_3 + H_2O$$

When a base is added

$$OH^- + H_2CO_3 \rightarrow HCO_3^- + H_2O$$

Blood pH

The ability of blood to maintain a pH of 7.40 is extremely important

Acidosis occurs when the pH of blood drops below 7.35

Alkalosis occurs when the blood pH rises above 7.45

Quiz 7B

Which is more acidic a pH of 6 or pH of 2?

High pH means
  (a) high hydronium ion concentration
  (b) low hydronium ion concentration.

Quiz 7B (cont.)

Low pH means
(a) high hydronium ion concentration
(b) low hydronium ion concentration

- How does a buffered system neutralize the effect of an added base?

Quiz 7B (cont.)

- Which of the following compounds can function as an antacid in the treatment of heartburn?

a) Sodium bicarbonate
b) Baking soda
c) Milk
d) Wine

Acid Rain

Any precipitation that has a measured pH of 5.60 or less is called **acid rain**.

Formed when sulfur dioxide (SO_2), and nitrogen monoxide (NO) combine with oxygen and moisture in the atmosphere and make sulfuric, nitric and nitrous acids.

Production of SO_2

Sulfur dioxide is a colorless gas formed as a by-product from the combustion of fossil fuels used in industrial processes, such as the production of iron, steel, utility factories, and crude oil processing.

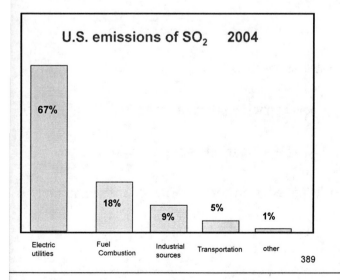

Production of NO

primarily produced from the combustion of fossil fuels.

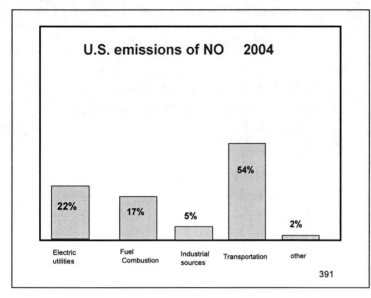

U.S. emissions of NO 2004

Formation of Acid Rain

sulfur dioxide oxidizes to form a sulfur trioxide (SO_3).

$$SO_{2\,(g)} + O_{2\,(g)} \rightarrow SO_{3\,(g)}$$

The sulfur trioxide then reacts with water forming sulfuric acid.

$$SO_{3\,(g)} + H_2O_{(l)} \rightarrow H_2SO_{4\,(aq)}$$

Formation of Acid Rain

nitrogen monoxide rises into the atmosphere reacting slowly with oxygen producing nitrogen dioxide (NO_2).

$$2NO_{(g)} + O_{2\,(g)} \rightarrow 2\,NO_{2\,(g)}$$

Formation of Acid Rain

Nitrogen dioxide then reacts with water to form nitrous acid (HNO_2) and nitric acid (HNO_3).

$$2\,NO_{2\,(g)} + H_2O_{(l)} \rightarrow$$

$$HNO_{2\,(aq)} + HNO_{3\,(aq)}$$

Chapter 8

Chemistry of Carbon
"The Organics"

Key Questions?

- What are fuels?
- How is energy produced from fuels?
- What are the major types of Hydrocarbons?
- What are isomers and why are they important?

Key Questions? Cont'd

- How do you refine petroleum?

- What is high octane gasoline?

- How and why is gasoline oxygenated?

- What about alternative fuels?

Hydrocarbons

Hydrocarbons are a class of compounds containing hydrogen and carbon only.

Four classes – Alkanes, Alkenes, Alkynes and Aromatics

Alkanes

Alkanes- Are saturated hydrocarbons with only carbon-carbon single bonds having the general formula.

$$C_nH_{2n+2}$$
$$\downarrow$$
$$C_2H_6 \quad \text{Ethane}$$

Condensed structural formula

Expanded structural formula

First Ten Straight Chain Hydrocarbons

n	Molecular Formula	Condensed structural formula	Name	Melting point ºC	Boiling point ºC
1	CH_4	CH_4	methane	-182	-162
2	C_2H_6	CH_3CH_3	ethane	-183	-89
3	C_3H_8	$CH_3CH_2CH_3$	propane	-190	-42
4	C_4H_{10}	$CH_3CH_2CH_2CH_3$	butane	-138	-1
5	C_5H_{12}	$CH_3CH_2CH_2CH_2CH_3$	pentane	-130	36
6	C_6H_{14}	$CH_3CH_2CH_2CH_2CH_2CH_3$	hexane	-95	69
7	C_7H_{16}	$CH_3CH_2CH_2CH_2CH_2CH_2CH_3$	heptane	-91	98
8	C_8H_{18}	$CH_3CH_2CH_2CH_2CH_2CH_2CH_2CH_3$	octane	-57	126
9	C_9H_{20}	$CH_3CH_2CH_2CH_2CH_2CH_2CH_2CH_2CH_3$	nonane	-51	151
10	$C_{10}H_{22}$	$CH_3CH_2CH_2CH_2CH_2CH_2CH_2CH_2CH_2CH_3$	decane	-30	174

Expanded Structural Formulas

use a single line to show every carbon-carbon as well as every carbon-hydrogen bond in the molecule.

For example the expanded molecular formula of propane would be:

Expanded Structural Formulas Cont'd

Propane C_3H_8

You try one!

Draw the expanded structural formula for hexane

Isomers of Alkanes

Isomers- Two or more compounds with the same molecular formula but different arrangements of atoms.

Isomer differ in one or more physical and chemical properties such as boiling point, color, solubility, reactivity and density.

Branched Chain Isomers

Structural isomers- Isomers that differ in the order in which the atoms are bonded together. (straight vs. branched chain)

Take C_4H_{10} as an example

```
                        C
                        |
   C-C-C-C            C-C-C
   Butane           2-methylpropane
```

Note: the hydrogen's have been omitted, but each carbon will have enough hydrogen's to make 4 bonds

Alkyl Groups

Alkyl groups- Alkanes with a hydrogen atom removed and are attached to a straight chain hydrocarbon.

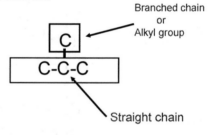

Names of Alkyl Groups

Alkyl groups are named from the parent hydrocarbon but the ending is changed from –ane to –yl.

Alkyl group has one carbon:
Parent = methane
Alkyl name = methyl

```
    C
    |
  C-C-C
```

Condensed formula $CH_3CH(CH_3)CH_3$

Naming Branched-Chain Alkanes

Locate the longest continues straight chain, this will be the parent name and number the carbons from left to right

Identify any alkyl groups attached to the longest chain and name them.

Identify which number carbon the alkyl groups are attached.

Write the carbon number, alkyl group name, then the parent name

Example of Naming

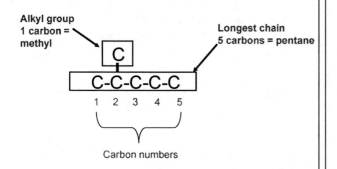

Name = **2-methylpentane**

Example of Naming

Name the following compound

$CH_3CH_2CH_2CH(CH_2CH_3)C(CH_3)_2CH_2CH_3$

Example of Naming Cont'd

Expanding gives

```
        C
        C   C
        |   |
    C-C-C-C-C-C-C
            |
            C
```

Then using the IUPAC rules

Example of Naming Cont'd

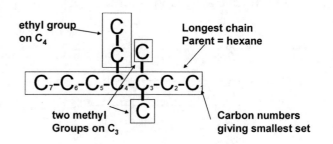

4-ethyl-3,3-dimethylheptane

Properties of Alkanes

C_1-C_4 are all gases
C_5-C_{17} are liquids
Above C_{18} are solids

Insoluble in water, less dense than water

They dissolve in many organic substances of similar polarities such as fats, oils, and waxes

Properties of Alkanes Cont'd

Alkanes are limited to only a few chemical reactions with the most important one being combustion

For this reason, alkanes are used mainly as fuels

Alkenes

Alkenes- hydrocarbons with one or more carbon-carbon double bonds.

Alkenes have the general formula C_nH_{2n}

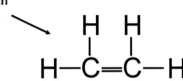

Naming Alkenes

Find the longest chain including the double bond and change the ending of the parent compound from –ane to –ene.

Number the carbon atoms as before, locate the carbon atom containing the double bond.

Name= number containing double bond, parent name ending in -ene

Naming Alkenes Cont'd

Name the following compound

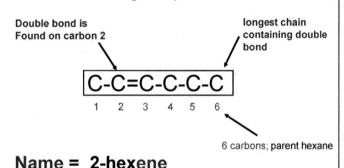

Double bond is Found on carbon 2

longest chain containing double bond

6 carbons; parent hexane

Name = 2-hexene

Structural Isomers of Alkenes

Compare 1-butene with 2-butene:

Both have same formula of: C_4H_8

C=C-C-C   C-C=C-C
1-butene   2-butene

*This small difference is responsible for very dissimilar properties

Alkynes

Hydrocarbons with one or more carbon-carbon triple bonds

General formula C_nH_{2n-2}

Name= just like alkenes except ending ends in -yne

C≡C—C—C

C_4H_6 = 1-butyne

Note: the hydrogen's have been removed, but each carbon would have enough hydrogen's to have 4 bonds and satisfy C_nH_{n-2} formula

Aromatic Hydrocarbons

Hydrocarbons containing one or more benzene rings

Benzene ring- formula C_6H_6

Aromatic Hydrocarbons Cont'd

Aromatic hydrocarbons are generally described as having strong and often pleasant odors, However, they are also known carcinogenic- cancer causing

Polymers

- Are huge molecules with molar masses ranging from thousands to millions.

- **Poly-** many
- **mer-** parts

- **Monomers-** small molecules that combine to form polymers

Polymers Cont'd

Polymers are classified by the reactions by which they were formed

Addition polymers- when all atoms in the monomers are incorporated into the polymer.

Polymers Cont'd

condensation polymer- Are formed by condensation reactions involving two molecules combining to make a larger molecule and a smaller molecule (usually water) thus the name condensation.

Addition Polymers

Polyethylene- produced from ethylene (monomer)

$$H_2C=CH_2 \longrightarrow \sim(H_2C-CH_2)n\sim$$

Common Addition Polymers

Formula	Monomer	Polymer	Uses
$H_2C=CH_2$	Ethylene	Polyethylene	Plastic bottles, films, toys
$H_2C=CHCH_3$	Propylene	Polypropylene	Indoor/outdoor carpet
$H_2C=CHCl$	Vinyl Chloride	PVC (polyvinyl chloride)	Plastic pipes, rain coats
$H_2C=CHCN$	Acrylonitrile	Polyacrylonitrile	Rugs, fabrics
$H_2C=CH(C_6H_5)$	Styrene	Polystyrene	Food and drink coolers

Condensation Polymers Cont'd

Polyester- is a polymer made from a molecule with 2 carboxylic acid and a molecule with 2 alcohols.

Nylon- is a polymer made from a di-acid and a di-amine

427

Quiz 8A

___ is the first member of the alkene series.

Name the following compound
$CH_3CH_2CH=CHCH_2CH_3$

The formula for the ethyl group is ____ .

Butane and 2-methylpropane are examples of _____.

428

Quiz 8A Cont'd

- What are polymers made from?
- PVC is made from what monomer?
- What polymer is used to make food and drink coolers?
- Monomer that undergo Addition polymerization must have what type of carbon-carbon bonds?

429

Quiz 8A Cont'd

- Nylon is an example of a (an) __ polymer.

- Polyester are formed by which type of reaction?

- What is the difference between addition and condensation reactions?

430

Alcohols

Alcohols- Organic compounds containing a hydroxyl (OH) functional group.

Functional Group- Atom or groups of atoms in a molecule that gives the substance a characteristic chemical behavior.

431

Alcohols Cont'd

Many alcohols are currently being tested for potential auto fuels, A common one (gasohol) is already at the pumps

Gasohol, (a mixture of 85% corn alcohol (ethanol) and 15% gasoline)

C-C-OH (ethanol)

432

Classes of Alcohols

Primary   Secondary   Tertiary

R = any alkyl group, and are not necessarily the same

Important Alcohols

Formula	BP (°C)	Synthetic name	Common name	use
CH₃OH	65	Methanol	Methyl alcohol	Fuel additives
CH₃CH₂OH	76	Ethanol	Ethyl alcohol	Beverages, fuel additives, solvents
CH₃CH₂CH₂OH	97	1-propanol	Propyl alcohol	Solvents
CH₃CH(OH)CH₃	82	2-propanol	Isopropyl alcohol	Rubbing alcohol
CH₂(OH)CH₂(OH)	198	1,2-ethanediol	Ethylene glycol	Antifreeze
CH₂(OH)CH₂(OH) CH₂(OH)	290	1,2,3-ethanediol	Glycerol	Moisturizers

Oxidation Products of Alcohols

Aldehydes- Organic compounds containing a –CHO functional group.

Carboxylic acids- Organic compounds containing a –COOH functional group

Ketones- Organic compounds containing a –C=O (carbonyl) functional group between two carbon atoms.

Oxidation of a Primary Alcohol

Primary alcohol → Aldehyde + Carboxylic acid

Oxidation of a Secondary Alcohol

Primary alcohol → Ketone

Aldehydes, Ketones and Carboxylic acids

Aldehyde   Ketone   Carboxylic acid

Formaldehyde — Embalming fluid

Acetone — Fingernail polish remover

Acetic acid — Vinegar

Carboxylic Acids

Are prepared from the oxidation of alcohols and aldehydes.

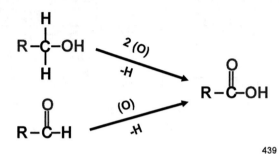

Common Carboxylic Acids

Formula	Common name	Synthetic name (Acid)	Where found
HCOOH	Formic acid	Methanoic	Ant stings
CH₃COOH	Acetic acid	Ethanoic	Vinegar
CH₃CH₂COOH	Propionic acid	Propanoic	Swiss cheese
CH₃(CH₂)₂COOH	Butyric Acid	Butanoic	Rancid butter
CH₃(CH₂)₃COOH	Valeric Acid	Pentanoic	Cow manure

Ethers

Organic compounds with the general formula R—O—R where

R= any alkyl group which may be the same or different.

Very common ether is diethyl ether once used as an anesthetic

CH₃CH₂-O-CH₂CH₃

Quiz 8B

- Name the simplest aldehyde?

- Which carboxylic acid is found in vinegar?

- What organic compound is in fingernail polish remover?

Fuels

Fuels are reduced forms of matter which burn easily in the presence of oxygen producing large quantities of **heat**.

Heat is a form of energy which can be used to do work.

Energy From Fuels

The energy is stored in the chemical bonds of the fuels.

This energy is released during the oxidation reaction (**combustion**) of fuels.

Energy From Fuels Cont'd

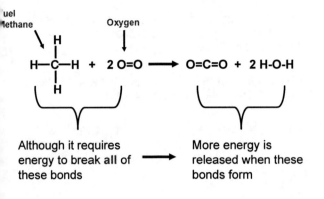

Although it requires energy to break all of these bonds → More energy is released when these bonds form

Energy values from fuels

Substance	Heat (kcal/g)
Hydrogen	34.0
Natural gas (methane)	11.7
Gasoline	11.5
Coal	7.4
wood	4.3

Sources of Energy in the United States

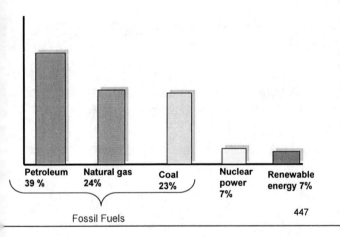

Petroleum 39% | Natural gas 24% | Coal 23% | Nuclear power 7% | Renewable energy 7%

Fossil Fuels

7.0 % Renewable Energy

Energy source	% Usage	Energy source	% usage
Solar	1.0 %	Waste	8.0 %
Wind	1.0 %	Wood	38.0 %
Alcohol	2.0 %	Hydroelectric	46.0 %
Geothermal	5.0 %		

Quiz 8C

All combustions of fossil fuels give off energy. a) True   b) False

Hydrocarbons react with ___ to produce CO_2 and _____ .

Name the following compound.
  $CH_3CH_2CH(C_2H_5)CH_2CH_2CH_3$

Quiz 8C Cont'd

Which fossil fuel furnishes the most heat energy per gram?
  a) Coal   b) Petroleum   c) Natural gas

Which fuel furnishes the most heat energy per gram?
  a) Natural gas   b) Hydrogen   c) Coal

Petroleum

Crude oil is a complex mixture of thousands of hydrocarbons, of which one is gasoline.

In order to separate the gasoline as well as the other hydrocarbons from one another the crude oil is refined

Refining process

Refining or separating hydrocarbon components from crude oil begins with a process call fractional distillation

Fractional distillation– Separation of a mixture into fractions that differ in boiling points.

Petroleum fractions– Mixtures of hundreds of hydrocarbons with boiling points in a certain range that are obtained by fractional distillation of petroleum.

Hydrocarbon fractions

Fraction	Size range	BP range °C	Uses
Gas	$C_1 - C_4$	0-30	Gas
Straight run gas	$C_5 - C_{12}$	30-200	Motor fuel
Kerosene	$C_{12} - C_{16}$	180-300	Jet fuel
Gas-oil	$C_{16} - C_{18}$	300-350	Diesel Fuel
Lubricants	$C_{18} - C_{20}$	Over 350	Lubricating oil
Paraffin wax	$C_{20} - C_{40}$	Low melting solids	Candles, wax
Asphalt	above C_{40}	residues	Roads, roofing

Hydrocarbon fractions

a barrel of crude oil contains only 35% gasoline

The demand for this fuel requires that over 50% of the crude oil be converted into gasoline.

Petroleum cont'd

Catalytic Cracking Process
Process by which larger kerosene fractions are converted into hydrocarbons in the straight-run gasoline range.

Straight run gasoline- straight-chain hydrocarbons, which burn too rapidly and cause uncontrolled explosions of the fuel which is characterized by a "knocking" or a "pinging" sound in the engine

Octane Rating

is an arbitrary scale for rating the relative knocking tendencies of gasolines. Heptane (C_7H_{16}), a straight-chain hydrocarbon, has been assigned an octane rating of 0 because it knocks considerably when used as auto fuel.

Octane Rating Cont'd

- On the other hand 2,24-trimethylpentane (CH$_3$C(CH$_3$)$_2$CH$_2$CH(CH$_3$)CH$_3$), burns with little or no knocking and is assigned an octane rating of 100

- All other gasolines are then measured by this scale

Petroleum cont'd

Catalytic Re-Forming

Process that increases octane rating of straight-run gasoline by converting straight-chain hydrocarbons to branched-chain hydrocarbons and aromatics.

Catalytic Cracking Process

Many organic chemicals used in the chemical industry are obtained from fossil fuels by catalytic cracking and fractional distillation techniques.

Catalytic Cracking

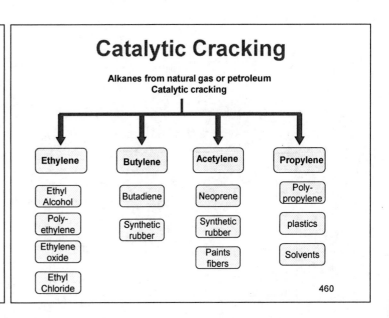

Fractional Distillation

Aromatics from petroleum or coal
Fractional distillation

- Benzene
  - styrene
  - Poly-styrene
  - Phenol
- Toluene
  - TNT
  - Aspirin
  - Plastics
- Xylene
  - Resins
  - Paints
  - Solvents

Organics Produced in the U.S.

Name	Produced	Uses
Ethylene	Cracking oil and natural gas	Plastics, fibers and solvents
Propylene	Cracking oil and oil products	Plastics, fibers and solvents
Urea	Reaction NH$_3$ and CO$_2$ under pressure	Fertilizer, animal feeds, adhesives
Styrene	Dehydration of ethylbenzene	Polymers, polyester

Quiz 8D

The fractions of petroleum are separated by
_____.

The principal component in natural gas is
_____ .

Quiz 8D cont.

The _____ process is used to produce branched-chain and aromatic hydrocarbons from straight-chain hydrocarbons.

The _____process is used in refining petroleum to convert molecules in the higher boiling fractions to molecules in the gasoline fraction.